Universal Navigation on Smartphones

Hassan A. Karimi

Universal Navigation
on Smartphones

Springer

Hassan A. Karimi
Geoinformatics Laboratory
University of Pittsburgh
School of Information Sciences
135 N. Bellefield Ave.
Pittsburgh Pennsylvania 15260
USA
hkarimi@sis.pitt.edu

ISBN 978-1-4899-9926-9   ISBN 978-1-4419-7741-0 (eBook)
DOI 10.1007/978-1-4419-7741-0
Springer New York Dordrecht Heidelberg London

Printed on acid-free paper

Springer is part of Springer Science+Business Media (www.springer.com)

Preface

Have you ever wondered why navigation systems have become so popular over the past several years? Do you know why navigation services are becoming common-place on mobile devices, particularly smartphones? The answer to these questions is that with advanced technologies, no one has to get lost while travelling from one location to another.

Mobility, or navigation, is a daily activity for all people everywhere, regardless of geographic location, mode of travel, time and duration of travel, travel needs, or travel purposes. Navigation can be a daunting task as transportation networks have become very large and complex over time. Large transportation networks mean increased ambiguities and difficult decisions to be made while in transit. Navigation could be further exacerbated when travelers are unfamiliar with their travelling environments. Ambiguities and navigation decisions, which may involve the whereabouts of an individual at any time during a trip or the step-by-step directions to be followed during a trip, have produced a need for navigation devices to provide navigation assistance.

Devices to assist with navigation have been around for decades and over that period, a wide range of technologies such as geo-positioning sensors, mobile computing, and wireless communications have evolved. Of these technologies, the Global Positioning System (GPS) mainly due to its pervasive nature, has had the most impact on navigation systems and services. This impact has been so profound that GPS has become synonymous with navigation, while in reality, it is only one of several components within a navigation system. Today, there are devices for out-door navigation assistance such as in-car navigation systems (offered as an extra in some vehicles), portable navigation systems (mobile devices assisting drivers), and navigation services (provided on cell phones and smartphones). While early navigation systems and services provided navigation assistance only for those in vehicles, there are emerging outdoor navigation systems and services to assist pedestrians with their navigation needs. In addition, we have seen the introduction of navigation systems and services to assist people in indoor wayfinding. While navigation systems and services for outdoors and indoors are conceptually and operationally similar, they are dissimilar with respect to the type of technologies and map data-bases they need to operate. For example, outdoor navigation systems and services are predominantly based on GPS for positioning cars and people, but GPS is not available or effective for indoor navigation systems and services; instead WiFi, one common geo-positioning sensor for indoor navigation, is used.

In this book, the current trends and future directions in navigation technology aimed at providing navigation assistance are discussed and analyzed. The book contains eight chapters covering topics on and related to navigation technology and navigation trends, research directions, issues, and challenges.

Chapter 1 discusses navigation technology from its early development to contemporary systems and services. The chapter starts with a historical perspective, providing an overview of the evolution of navigation technology, and discusses the specific characteristics of navigation environments, technological advances, and the challenges facing navigation technology in terms of indoor/outdoor travel and information sources. In particular, the similarities and differences between outdoor and indoor navigation systems and services are compared and analyzed. The chapter also discusses the shortcomings of current navigation systems and services, and briefly presents the concept of universal navigation and its potential to overcome those shortcomings.

Chapter 2 discusses outdoor navigation in detail. It starts by describing the information flow in outdoor navigation systems and highlighting the main components of navigation systems including geo-positioning, map matching, geocoding, mapping, routing, and directions. The three main technologies, i.e., geo-positioning, wireless communication, and database, supporting outdoor navigation systems are examined. This is followed by descriptions of the typical functions performed in these systems. The chapter ends with a discussion of usability in outdoor navigation systems.

Chapter 3 provides an in-depth exploration of indoor navigation, including the three primary technologies, i.e., geo-positioning, wireless communication, and database, of indoor navigation systems. This chapter goes on to describe the typical functions performed in these systems and concludes with a discussion of usability in indoor navigation systems.

Chapter 4 presents the concept and technologies of Universal NAVIgation Technology (UNAVIT), which can provide navigation assistance anywhere, anytime, and for any user. The chapter discusses the UNAVIT's capabilities, as well as an ontology. The features of UNAVIT, assistance for travel anywhere, anytime, by any user, in any mode of travel automatically and adaptively, are discussed. A possible architecture for UNAVIT is described, outlining the various components of such a system. The chapter also looks at information flow in UNAVIT for smartphones (Android and iPhone) and Web Mapping Services (e.g., Google Maps, Bing Maps, Yahoo Maps).

Chapter 5 details the anywhere feature of UNAVIT. It begins with an ontology highlighting the concepts of anywhere navigation and the relationships among those concepts. Then, the different categories of anywhere navigation are analyzed: indoor navigation, outdoor navigation, indoor-outdoor navigation, outdoor-indoor navigation, and indoor-outdoor-indoor navigation. Two algorithms, one for navigation from outdoor to indoor and another for navigation from indoor to outdoor, are described and analyzed.

Chapter 6 examines the anytime feature of UNAVIT, discussing how navigation assistance could be affected by time of day, day of week, and the season. An ontology to highlight the concepts in anytime navigation and the relationships among them is presented. The chapter also describes an algorithm that addresses navigation assistance based on different times and situations.

Chapter 7 discusses the anyuser feature of UNAVIT at a deeper level. The categories of users (general population, mobility-impaired, visually-impaired, cognitively-impaired, elderly) are discussed. The chapter presents an ontology to highlight the different needs and preferences of each category of users and describes an algorithm to address the needs and preferences of each of those users.

Chapter 8 presents social navigation networks (SoNavNets) as a new method for providing navigation assistance. The chapter starts by defining location-based social networks, a new approach to providing location-based services through social networks. An ontology for SoNavNets, along with algorithms for request and recommendation, are outlined. To better understand the components and capabilities of this new approach, a social navigation network (SoNavNet) system, developed by the author in the Geoinformatics Laboratory of the School of Information Sciences at the University of Pittsburgh, is presented. The chapter discusses and analyzes the differences between the model-centric navigation assistance (the current computing approach) and the experience-centric navigation assistance (the SoNavNet approach) where members share navigation experiences with others.

Table of Contents

Chapter 1
Introduction to Navigation

1.1 Introduction

Navigation, in the context of this book, is defined as movement of people from one location to another. Navigation assistance is a means by which people can be provided with information about their navigation needs and preferences. Modern navigation technology has become an attractive means for assisting people with their navigation needs and preferences, and since its inception (mid 1980s) the demand for it has sharply increased. In this chapter, an overview of navigation technology, its past, current trends, and future directions are discussed.

In this book, navigation both in outdoors and indoors is discussed. However, since these two navigation environments, i.e., navigation in outdoors and navigation in indoors, share common characteristics and have differences, some unique aspects of outdoor and indoor navigation are overviewed in this chapter. Moreover, considering that early navigation technology was built for outdoor use only, an overview of outdoor navigation technology is followed by an overview of indoor navigation technology.

Before reviewing navigation technology, its characteristics and generations, it is important to emphasize that the need for navigation assistance is not something new. For centuries, people have been trying to develop devices and tools that assist them in navigating new and unfamiliar environments. Although there are different perspectives in categorizing the developments of navigation devices and tools throughout the history, two general categories are mechanical and digital. Up until the advent of computers, most navigation devices were of mechanical type in that they operated based on the principles of physical mechanics. Perhaps, one of the earliest mechanical navigation devices reported is the South Point Chariot (Figure 1.1*). The chariot was invented around 2600 BC in China and resembles a two-wheeled vehicle where through measuring the differences between the rotations of the wheels an object connected to them would always point to the same direction. Figure 1.2* shows a reconstructed version of the South Pointing Chariot.

---

* http://en.wikipedia.org/wiki/South_Pointing_Chariot

H. A. Karimi, *Universal Navigation on Smartphones,*
DOI 10.1007/978-1-4419-7741-0_1, © Springer Science+Business Media, LLC 2011

Fig. 1.1 South pointing
chariot

Fig. 1.2 A reconstructed
south pointing chariot

The advent of computers, among other technologies, paved the way for the development of digital navigation devices and tools, those that are based on the principles of electronics. In this book, digital navigation devices are focused, which became popular with the debut of the Global Positioning System (GPS) in the mid 1980s and its promise of being a novel technology for many applications including data collection and navigation. GPS has had a tremendous impact on the development of modern navigation technology. The impact has been in threefold. One is that GPS has allowed ubiquitous, anywhere, anytime, positioning, with a level of accuracy and reliability suitable for a wide range of land-based navigation activities. Second is that as GPS has been used in numerous existing and new applications, people have realized its benefits, especially for land-based navigations. Third is that GPS receivers have become very inexpensive, making modern navigation devices affordable and commonplace.

Given the importance of geo-positioning technologies in navigation applications and the widespread existence of GPS in navigation devices, navigation technology is often referred to as "GPS Systems" or simply "GPS". This is just like referring to computers as "CPUs" because a CPU is the most important component of a computer. But this is not correct, because just as a computer with only a CPU and without the other components (e.g., memory) cannot handle computing, it is difficult to navigate with a navigation device equipped only with a GPS receiver and without the other components (e.g., map matching).

Geospatial Information Systems (GIS) is another technology, which was under refinement and improvement in parallel in the 1980s, that played a key role in the development of modern navigation technology. GIS was recognized as the underlying technology where its data (spatial data on which GIS operations are based constitutes the core of the data in the database of a navigation system) and its functions (e.g., routing) were needed to be integrated and utilized with GPS data. The process was conceptually simple, GPS provided the real-time position of people and objects where GIS used GPS data, along with other spatial and non-spatial data, and specific functions to provide navigation assistance. Given the nature of this integration, i.e., GPS data in a GIS for data processing and computation, navigation devices can be considered as specialized GIS applications. In short, the synergy between GPS and GIS gave birth to modern navigation technology.

Figure 1.3 shows the overall flow of information in today's navigation technology. The purpose of the figure is to show the flow of information independent of any

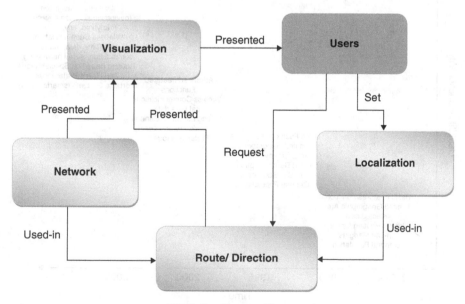

Navigation Concept Diagram

Fig. 1.3 Flow of information in navigation systems

specific technology or any specific technique. As shown in this figure, upon initiation of a user for navigation assistance requesting a route, the localization component that determines user's current location is invoked; this is accomplished through any one or a combination of geo-positioning sensors. User's request for a route/direction, which is typically a preferred route between user's current location and a desired destination, is computed by the route/direction component. These two major activities require network data (e.g., road segments), among other data, which is the core of data in modern navigation technology. Final computed results (current location, route, direction) are overlaid and presented (visualized) on maps. Note that Figure 1.3 only illustrates a high-level flow of information in navigation systems. Chapters 2 and 3 provide much more in-depth details about information flow, technologies, and components in navigation systems in outdoors and indoors, respectively.

1.2 Outdoor Navigation

To give an overview of the development of outdoor navigation technology, the evolution of outdoor navigation technology is divided into four generations. Figure 1.4 shows the four generations of navigation technology for outdoors and their timelines.

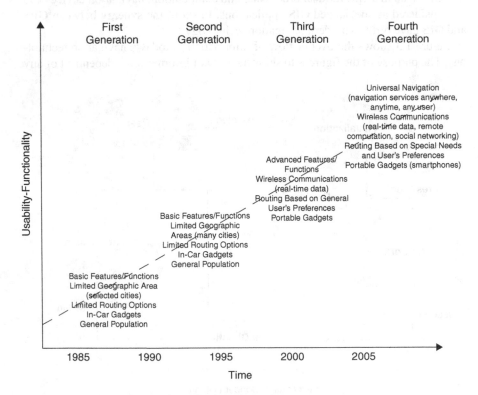

Fig. 1.4 Generations of navigation technology for outdoors

The first generation of navigation technology (~1985 - ~1995) offered basic features and functions; was available for very limited geographic areas (selected cities); offered limited routing options (mainly shortest route); was only available as in-car gadgets (they were installed as luxury gadgets by the automobile manufactures on selected cars); and provided navigation assistance to the general population.

The second generation of navigation technology (~1995 - ~2000) offered basic features and functions. However, as advanced techniques were developed and feedbacks from users were incorporated, the features and functions in the second generation were the improved version of those in the first generation. The second generation of navigation technology was made available for various geographic areas (many cities); supported limited routing options; was available as portable devices; and provided navigation assistance to the general population.

The third generation of navigation technology (~2000 - ~2005) offered advanced features and functions; provided (optional) wireless connection (primarily to obtain real-time data); was available on mobile devices and personal digital assistants (PDAs); and offered routing options that met the preferences by the general population. Figure 1.5 shows an example of outdoor navigation systems in the third generation.

The fourth generation of navigation technology (~2005 -), which is the current trend, offers navigation services with a variety of new features addressing personalized navigation needs and preferences anywhere, anytime and for any user. Navigation technology in each of the first three generations can be characterized as generic and system-oriented, assisting with general navigation activities, and are either in-car navigation systems or portable navigation devices. The fourth generation of navigation technology is characterized as personalized and service-oriented, assisting with navigation activities at individual level, where services are provided by navigation service providers.

System-oriented navigation assistance and service-oriented navigation assistance can be distinguished by data storage, computation, and communication. Navigation systems are stand-alone devices that can provide navigation assistance without connection to external services where they contain all the required data and can provide all the required computations. Navigation services are provided through

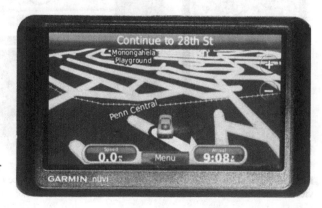

Fig. 1.5 Example of outdoor navigation systems in the third generation

Table 1.1 System-oriented and service-oriented navigation assistance

Type	Data	Computation	Communi-cation	Comments
System (stand-alone devices)	All data stored on device	All computations performed on device	Not needed	Data and computation update: time consuming and manual Cost: one time
Service (bundled with smartphone services)	Option 1: All data stored on remote servers Option 2: Most data stored on remote servers	Option 1: All computations performed at remote servers Option 2: Most computations performed at remote servers	Needed	Data and computation update: automatic and transparent Cost: per usage

mobile devices (e.g., smartphones) where most of the required data and most of the required computations are provided through remote servers maintained by service providers. Table 1.1 shows the differences between system-oriented navigation assistance and service-oriented navigation assistance. It should be noted that due to the compactness, portability, and affordability features of navigation services, and the fact that navigation is one among several services provided on smartphones, the demand for navigation services is rapidly growing.

Figures 1.6 (a) and 1.6 (b) show example navigation services on iPhone and on Android, two predominant smartphones currently in the market.

Fig. 1.6 Navigation services on smartphones. **a** Navigation serivce on iPhone. **b** Navigation service on Android.

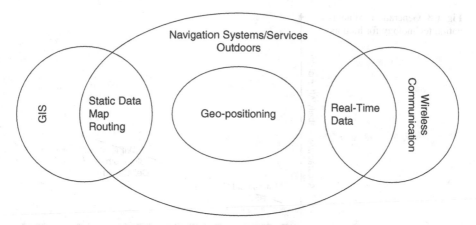

Fig. 1.7 Key technologies in navigation systems/services for outdoors

Figure 1.7 shows the different technologies that play key roles in navigation systems/services for outdoors. Geo-positioning is at the heart of navigation systems/services for outdoors in that most navigation activities are dependent upon position information provided by geo-positioning sensors, primarily Global Satellite Navigation System (GNSS). GIS contributes by providing core static data including maps and core navigation functions such as routing in navigation systems/services. Wireless communication contributes by providing real-time (dynamic) data, such as traffic, in navigation systems/services for outdoors. As shown in this figure, while wireless communication can be used as a geo-positioning sensor (e.g., WiFi), its geo-positioning role is not as dominant as GPS is for navigation in outdoors.

1.3 Indoor Navigation

Compared to outdoor navigation systems/services, the evolution of indoor navigation systems/services has a much shorter time span. This is perhaps due to the fact that navigation in outdoors is much more complex than navigation in indoors. Navigation in outdoors imposes certain constraints, such as real-time decision making (especially when driving), requiring solutions to navigation problems in a much larger space (geographic area) and finding optimal routes from a very large solution space (number of options). For example, a trip may require an optimal route among many possible options between a pair of origin and destination locations in a large city, and while enroute to the destination a new route may be needed due to change in weather or traffic or occurrence of accidents.

The evolution of indoor navigation technology can be divided into two generations as shown in Figure 1.8.

As shown in Figure 1.8, in the first generation of indoor navigation technology, which debuted in the mid-1990s, only a few geo-positioning sensors were available.

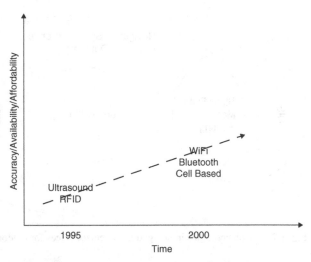

Fig. 1.8 Generations of navigation technology for indoors

In general, geo-positioning sensors for indoor navigation were scarce and unaffordable. The second generation of indoor navigation technology, which debuted around early 2000s, has enjoyed new geo-positioning sensors and techniques which offer improved accuracy and are widely available and affordable.

It is important to note that compared to navigation in outdoors, where both navigation systems and services could be utilized for navigation assistance, navigation assistance in indoors is more meaningful and practical through navigation services on ubiquitous devices such as cell phones (increasingly smartphones). While it is common and practical that people requiring navigation assistance for driving, biking, or walking in outdoors utilize navigation systems or services provided on mobile devices, it is hard to imagine that people would be walking within a building with specialized mobile devices to find a room in the building. On the other hand, it is conceivable to imagine that people would be provided with navigation assistance in indoors through navigation services on smartphones as they are becoming commonplace alleviating the need to carry extra devices for the purpose of navigation.

Figure 1.9 shows an example of current indoor navigation technology.

Figure 1.10 shows the different technologies that play key roles in navigation systems/services for indoors. Like navigation in outdoors, geo-positioning is at the heart of navigation technology for indoors in that most navigation activities are dependent upon position information provided by geo-positioning sensors. However, unlike navigation systems/services for outdoors which are predominantly based on GNSS (e.g., GPS) for geo-positioning, GPS does not play a significant role in navigation systems/services for indoors. Instead, wireless communication (e.g., RFID, WiFi), whose role in outdoor navigation is primarily for receiving real-time (dynamic) information, is the predominant geo-positioning sensor in indoors. For the reason that indoor navigation is not affected by environmental factors, such as weather condition, there is really little need to receiving real-time data by wireless communication. Computer-Aided Design (CAD) and Building Information Model

Fig. 1.9 Example of current indoor navigation technology

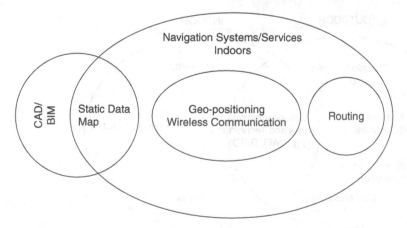

Fig. 1.10 Key technologies in navigation systems/services for indoors

(BIM) contribute by providing core static data including maps. As shown in this figure, navigation functions, such as routing, could be included as separate modules into navigation systems/services for indoors.

1.4   Outdoor vs. Indoor Navigation

Figure 1.11 shows the technologies and techniques that are used for both outdoor navigation and indoor navigation. As shown in this figure, the only overlap between the two navigation environments is geo-positioning sensors and techniques (e.g., RFID, WiFi,) that are applicable in both environments.

   To better understand the differences and the similarities between mobility in outdoors and in indoors, they are compared and analyzed with respect to navigation and routing. Table 1.2 shows the requirements of navigation and Table 1.3 shows the requirements of routing in outdoors and indoors, respectively.

   As shown in Table 1.3, outdoor navigation systems/services must provide navigation assistance for different modes of travel including driving, walking, biking, riding wheelchairs, whereas indoor navigation systems/services only need to assist in walking and riding wheelchairs. The data for outdoor navigation systems/services comes primarily from GIS databases which include road and/or sidewalk segments (with such attribute data as type, width, length) and road and/or sidewalk networks. The data for indoor navigation systems/services comes primarily from CADs which primarily include hallway segments (with such attribute data as width, length) and hallway networks. Outdoor navigation systems/services contain data, and appropriate functions to process them, at different scales in order to be able to provide navigation assistance in as large scale as a neighborhood to as small scale as a country. Indoor navigation systems/services contain data, and appropri-

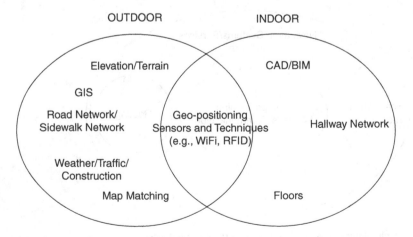

Fig. 1.11 Technologies and techniques for outdoor and indoor navigation systems/services

Table 1.2 Navigation requirements in outdoors and indoors

Navigation Environment	Mode of Travel	Database	Scale	POIs	Factors Impacting Navigation	Spatial Complexity
Outdoor	Driving; Walking; Biking; Riding wheelchairs	GIS: road/sidewalk segment, road type, width, length, speed limit Network: road/sidewalk	Neighborhood, city, county, state, country	House, restaurant, gas station, hotel, office building	Weather, traffic, accident, construction	Horizontal: plain to hillside
Indoor	Walking; Riding wheelchairs	CAD/BIM: hallway segment, width, length Network: hallway	Building	Room, restroom	Refurbishment	Horizontal/Vertical: floor to multi-floors

Table 1.3 Routing requirements in outdoors and indoors

Routing Environment	Mode of Travel	Network	Optimal Criteria	Scale	Factors Impacting Route Selection	Problem Space	Computation Complexity
Outdoor	Driving, Walking; Biking; Riding wheelchairs	Road, sidewalk	Shortest; Fastest; Least intersections; Least turns; Least traffic; Most scenic	Neighborhood, city, county, state, country	Weather, traffic, accident, construction	Large to very large networks	Low to very high
Indoor	Walking; Riding wheelchairs	Hallway	Shortest	Building	Refurbishment	Small network	Low

ate functions to process them, about buildings, usually at a fixed scale, to provide navigation assistance within the buildings. While outdoor navigation systems/services contain a large variety of points of interest (POIs) such as house, restaurant, gas station, hotel, and office building, indoor navigation systems/services contain limited POIs such as office (room), restroom, and exit. Outdoor navigation can be impacted by such factors as weather, traffic, and accident, while such factors are of no concerns in indoor navigation.

As shown in Table 1.3, routing in outdoor must address a wide range of criteria such as shortest, fastest, least intersections, least turns, least traffic, most scenic, among others, whereas the main routing criterion in indoor is shortest distance. Routing in outdoor typically involves large size road/sidewalk networks potentially causing computation complexity, whereas routing in indoor typically involves small size networks with low computation complexity.

In this book, the characteristics, technologies, techniques, and issues of the fourth generation of outdoor navigation technology and those of the second generation of indoor navigation technology are discussed and analyzed. These generations are called universal navigation which is service-oriented providing personalized navigation assistance anywhere, anytime, and for any user. Smartphones are well suited for universal navigation as they support multiple geo-positioning sensors, to obtain user's position anywhere, feature multiple-wireless communication systems, to obtain real-time data (in outdoors) and user's position (in indoors), can transfer computations to remote servers, and facilitate social networking. With such advanced technologies, smartphones can provide navigation assistance and routing options that meet the needs and preferences of all users including the general population and the individuals with special needs, e.g., physically, visually, cognitively impaired. Universal navigation (fourth generation of outdoor navigation technology and second generation of indoor navigation technology) is currently taking place and with the prevalence of smartphones, it is anticipated that both outdoor navigation and indoor navigation services converge into one platform where users can be anywhere (outdoor or indoor) and be able to navigate seamlessly between outdoor and indoor environments.

1.5 Shortcomings

Despite much upward progress in navigation systems/services with respect to usability, functionality, and affordability, the high demand for them, and their availability in many places around the world, they have shortcomings. Table 1.4 summarizes the shortcomings of current navigation technology. The questions in the table highlight some of the shortcomings, among others, that are observed by individuals based on various factors including purpose of trip, personal preferences, and user interfaces. The fourth generation of outdoor navigation technology and the second generation of indoor navigation technology are intended to address and overcome these shortcomings.

Table 1.4 Shortcomings of current navigation technology

Questions	Comments
Can modern navigation technology meet the specific needs of all individuals including those with physical (mobility) and sensory (vision, hearing), and cognitive impairments?	There are experimental projects which address some of these needs. No commercially available navigation systems currently can address the range of different needs.
Is modern navigation technology adaptable to users with different cognitive abilities, levels of computing knowledge and exposure to technology?	While navigation technology has improved, current navigation systems still require some level of knowledge and comfort with computer technology for their effective utilization.
Can modern navigation technology provide navigation assistance seamlessly between indoors and outdoors?	There are very few experimental projects that have this feature. Commercially available navigation systems are either for outdoors or for indoors, not both.
Can modern navigation technology support users with navigation assistance suitable for various situations, such as difficult versus easy routes, day versus night, routes on snowy days versus sunny days?	Some current navigation systems address some of these issues.
Can modern navigation technology provide individuals with proper information seamlessly based on the context?	There are experimental projects that are context aware. Commercially available navigation systems are not context aware.
Can modern navigation technology seamlessly adjust to different modes of travel, such as driving cars, walking, riding bicycles, riding wheelchairs or Segways?	There are few current navigation systems capable of realizing and adjusting more than one mode of travel but not all possible modes.
Can modern navigation technology effectively be used in different countries with dissimilar policies and cultures?	Current navigation systems are not able to address these issues.

1.6   Universal Navigation

By universal navigation, a navigation framework that facilitates personalized navigation assistance anywhere, anytime, and for any user using smartphones is referred. This new framework is called UNiversal NAVIgation Technology (UNAVIT), which is the foundation of both the fourth generation of outdoor navigation technology and the second generation of indoor navigation technology. UNAVIT is service-oriented providing navigation assistance on demand regardless of the geographic area in which the assistance is requested and whether the assistance is needed for outdoor or indoor. Unlike current navigation systems/services that are one-size-fits-all and cannot handle seamless transition from indoor to outdoor or from outdoor to indoor, are not adaptable to time-specific navigation (e.g., navigation during non-rush hours vs. rush hours), address primarily needs and preferences of the general population, UNAVIT is intended to support location-independent, time-independent, and user-independent navigation assistance.

One new approach in UNAVIT, which is feasible and viable for addressing the universality of navigation assistance both in outdoors and indoors, is navigation through social networking. This is because through social networking navigation experiences of members where there is element of "trust" are shared which sometimes in some places are available where conventional (i.e., computing) approach is not. For this, UNAVIT, as the latest generation of navigation technology, which is based on a synergy between "computing", through the conventional approach, and "people", through the social networking approach, will address the anywhere, anytime, and any user features of navigation.

Usability is another important feature of UNAVIT. UNAVIT is expected to assist the users with various navigation needs and preferences including those required by the general population and those required by the special needs population. Figure 1.12 shows the three levels of usability in navigation. The level at the bottom reflects navigation usability for the general population. This level of usability was the main focus of the first, second, and third generations of outdoor navigation technology and the first generation of indoor navigation technology. The level in the middle reflects the needs of various groups such as physically and cognitively impaired. The level at the top reflects the preferences of the general population, the

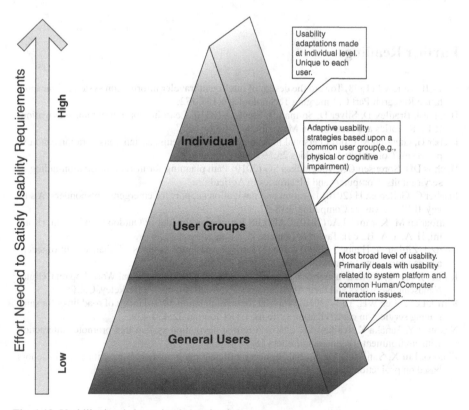

Fig. 1.12 Usability levels in navigation technology

requirements by the special needs groups plus the preferences of individuals. This top level usability is also referred to as navigation personalization and together with the middle level are the focus of the fourth generation of outdoor navigation technology and of the second generation of indoor navigation technology.

1.7 Summary

This chapter gives an overview of navigation and navigation assistance in general. Indoor and outdoor navigation, their characteristics and differences are discussed. The evolution of outdoor navigation technology is divided into four generations and the evolution of indoor navigation technology is divided into two generations. Shortcomings of earlier generations of outdoor and indoor navigation technologies, where they provided limited functionality and allowed selected options, are discussed. The trend in the fourth generation of outdoor navigation technology and in the second generation of indoor navigation technology is universality of navigation (UNAVIT) in that personalized navigation assistance can be provided anywhere, anytime, and for any users.

Further Readings

Adler JL, Blue VJ (1998) Toward the design of intelligent traveler information systems. Transportation Research Part C: Emerging Technologies 6:157-172.

Bagnell J, Bradley D, Silver D, Sofman B, Stentz A (2010) Learning for autonomous navigation. IEEE Robotics and Automation Magazine 17:74-84.

Bieber G, Giersich M (2001) Personal mobile navigation systems–design considerations and experiences. Computers & Graphics 25:563-570.

Bochtis DD, Sørensen CG, Vougioukas SG (2010) Path planning for in-field navigation-aiding of service units. Computers and Electronics in Agriculture.

Fischer C, Gellersen H (2010) Location and navigation support for emergency responders: A survey. IEEE Pervasive Computing 9:38-47.

Ghafourian M, Karimi, H.A. (2010) CAD/GIS Integration for Indoor/Outdoor Navigation. (Karimi, H. A. a. A. B., ed): Taylor & Francis.

Goetz J, Zittlau D, Happe J (2006) Advanced driver assistance systems - Enhancement of safety and comfort. AutoTechnology 6:34-38.

Karacs K (2010) Towards a mobile navigation device. In: 12th International Workshop on Cellular Nanoscale Networks and their Applications (CNNA 2010), pp 1-4 Berkeley, California.

Kim J, Lee J, Lee WH, Yu K (2010) Proposal for an inundation hazard index of road links for safer routing services in car navigation systems. ETRI Journal 32:430-439.

Nakatani Y, Tanaka K, Ichikawa K (2010) A tourist navigation system that promotes interaction with environment. Engineering Letters 18.

Zhao G, Liu X, Sun M-T, Ma X (2008) Energy-efficient geographic routing with virtual anchors based on projection distance. Computer Communications 31:2195-2204.

Chapter 2
Outdoor Navigation

2.1 Introduction

In this chapter, the technologies and techniques that are employed in outdoor navigation systems/services along with their features and users are discussed. While the first and second generations of outdoor navigation technology were limited to specific technologies (e.g., only a few geo-positioning sensors were feasible and/or cost effective) and to specific techniques (e.g., routing modules were capable of dealing only with single optimization criterion and there were very few criteria from which users could select), a wide range of technologies and techniques have been available for the third and fourth generations of outdoor navigation technology.

This chapter discusses existing navigation systems/services for outdoor navigation. Despite apparent distinctions among current outdoor navigation systems/services, they are all based on the same fundamental logic and support similar functions/modules. Figure 2.1 depicts the information flow and functions/modules in outdoor navigation systems/services.

As shown in Figure 2.1, in the first step user's current position is determined by: (a) obtaining position data through geo-positioning sensors and (b) applying a map matching algorithm using the obtained position data. It is common practice, in order to improve the accuracy, availability, and reliability of position data in outdoor navigation systems/services, to employ more than one geo-positioning sensors where the acquired position data at each epoch could be filtered (e.g., through a Kalman filter) to find the best position estimate. Once the position data is obtained (directly or filtered), it is input to the map matching algorithm where it uses a map database of the traveling area, which includes spatial and non-spatial data, to find: (a) the road/sidewalk segment on which the user is and (b) the precise location of the user on the segment.

Once user's current location on a road or sidewalk segment is determined, it is highlighted on the map and presented to the user. From this point on, the system/service is on tracking mode, and the user has options to search for POIs or request for optimal routes between pairs of addresses. Upon a route request, the system/service uses such routing criteria as shortest or fastest (often pre-determined by the user)

H. A. Karimi, *Universal Navigation on Smartphones*,
DOI 10.1007/978-1-4419-7741-0_2, © Springer Science+Business Media, LLC 2011

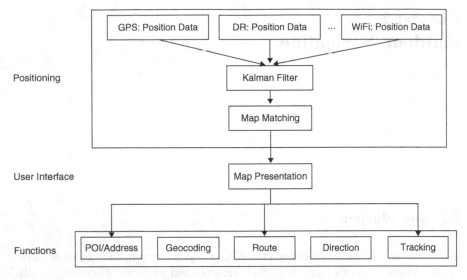

Fig. 2.1 Flow of information in outdoor navigation systems/services

to compute a route between current location, as the origin, and a given address, as the destination. Then the computed route is used to generate a set of turn-by-turn instructions which are presented to the user on the map and/or through voice. While enroute to the destination, the system/service continues obtaining user's location (using the same sequence of obtaining position data through geo-positioning sensors and applying the map matching algorithm) until the destination is reached. In case the user deviates from the computed route, a new route and a set of new directions are computed (this is commonly known as rerouting) by using the newly acquired position data (i.e., user's current location), as the origin, to the destination. This process continues until the user arrives at the destination.

2.2 Technologies

The main technologies used in outdoor navigation systems/services include geo-positioning, wireless communication, and database. The main modules in outdoor navigation systems/services are mapping, geocoding, map matching, routing and directions. Depending on several factors including application requirements and cost, navigation system/service vendors decide on the technologies and modules that are feasible and cost effective. However, nowadays navigation system/service vendors are putting more efforts into addressing the needs and preferences of various users in their navigation products. Examples of users' needs include routes which contain roads with a certain speed limit and roads with two or more lanes. Examples of users' preferences are routes with shortest distance, fastest travel time, simple directions, no congestion, scenic sites, and no tolls. Addressing these needs and preferences require the integration of different technologies and techniques in an outdoor navigation system/service. The technologies and techniques suitable for outdoor navigation systems/services are described in the reminder of this section.

2.2.1 Geo-positioning

Geo-positioning sensors provide position data in navigation systems/services. Today, there are various geo-positioning sensors that are employed in outdoor navigation systems/services and have been the subject of many books and papers where their characteristics and issues, especially accuracy, are discussed. Figure 2.2 shows possible levels of accuracy provided by various geo-positioning sensors that can be implemented in outdoor navigation systems/services. One way to distinguish between these geo-positioning sensors is to characterize the type of positioning technique provided by each sensor (i.e., absolute positioning or relative positioning). Absolute positioning is a technique to calculate the position of an object without using a point of reference. Relative positioning is a technique to calculate the position of an object by using a point of reference. For example, GPS is an absolute positioning technique and Differential-GPS (DGPS) is a relative positioning technique; another example is dead reckoning (e.g., differential odometer) which is a relative positioning technique.

In order to better understand the positioning needs of outdoor navigation systems/services, this section discusses the geo-positioning sensors that are suitable for outdoor navigation systems/services. The existing geo-positioning sensors suitable for outdoor navigation systems/services can be categorized based on three infrastructures: global, wide, and local (Table 2.1). The rationale for this infrastructure-based categorization is that, depending on the underlying infrastructure, each sensor has a different coverage.

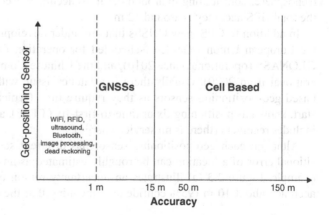

Fig. 2.2 Levels of accuracy provided by geo-positioning sensors

Table 2.1 Geo-positioning sensors suitable for outdoor navigation

Infra-structure	Coverage	Sensor	Accuracy	Latency*	Cost to User
Global	Anywhere on earth	Global Navigation Satellite Systems (e.g., GPS, Galileo, GLONASS, Compass)	Better than 15 m	Slow	Receiver
Wide	City, state, region	Cell based	50-150 m	Medium	Mobile device, service
Local	Campus, city	WiFi, RFID, ultrasound, Bluetooth, image processing, dead reckoning	Centimeters to meters	Fast	Device (HW/SW), service

*Latency: slow (minutes), medium (minute), fast (seconds)

Table 2.2 Sources of errors (in meters) in GPS

Satellite		Receiver		Environment	
Ephemeris	Clock	Measurement	Multipath	Ionosphere	Troposphere
2.5	2.0	1.0	1.0	5.0	0.5

Global infrastructure-based geo-positioning sensors cover everywhere on earth and typically provide better than 15 meters accuracy. GNSS is the best example of global infrastructure-based geo-positioning sensors. The United States NAV-STAR GPS, which is the only operational GNSS as of 2011, is the most prominent GNSS. The three sources of error in GPS are: satellite, receiver, and environment. Table 2.2 shows the sources of errors and the parameters under each source. Each parameter contributes a certain amount of error to the total accuracy in GPS. The two parameters under the satellite source are ephemeris, contributing to around 2.5 m accuracy, and clock, contributing to around 2.0 m accuracy. The two parameters under the receiver source are measurement, contributing to around 1.0 m accuracy, and multipath, contributing to around 1.0 m accuracy. The two parameters under the environment source are ionosphere, contributing to around 5.0 m accuracy, and troposphere, contributing to around 0.5 m to accuracy. Accumulating these errors, the total GPS accuracy is around 12 m.

In addition to GPS, new GNSSs that are under development or deployment are the European Union's Galileo (scheduled for operation from 2012), the Russian GLONASS (operational since 2010), and the Chinese Compass (planned to be operational from 2020). Usually there is a latency issue with global infrastructure-based geo-positioning sensors as they require time (typically within minutes) to start, known as positioning fix or time-to-first-fix (TTFF), and their cost to the user includes receivers (there is no service charge).

Although each geo-positioning sensor has different sources of errors, the positional error of a location can be roughly estimated as an uncertainty region. For example, Figure 2.3 (a) illustrates an uncertainty region of GNSS positions with accuracy about 10 m (95% confidence), meaning that the estimated point will be

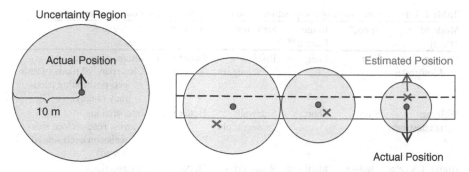

Fig. 2.3 Positional uncertainty. **a** Uncertainty region for an actual position. **b** Estimated positions and uncertainty regions for a moving object

located within 10 m of the actual location. Figure 2.3 (b) illustrates a moving point with different levels of positional accuracy; the cross marks are estimated positions of the vehicle obtained from GNSS while the circle marks are actual positions of the vehicle.

Wide infrastructure-based geo-positioning sensors cover a city, a state, a region and have accuracy ranging between 50 meters to 150 meters. Cell-based systems are the best example of the wide infrastructure-based geo-positioning sensors. The wide infrastructure-based geo-positioning sensors have medium (typically within a minute) latency and their cost to the user includes mobile devices and services.

Local infrastructure-based geo-positioning sensors cover a campus or a city and have an accuracy range of centimeters to meters. Examples of local infrastructure-based geo-positioning sensors are RFID, WiFi, dead reckoning, image processing, ultrasound, and Bluetooth. The local infrastructure-based geo-positioning sensors have the fastest start time (typically within seconds) and their cost to the user includes devices (hardware and software) and services.

Table 2.3 shows the factors that must be taken into consideration for choosing geo-positioning sensors appropriate for outdoor navigation. In the table, the navigation needs of users in outdoors based on their mode of travel (i.e., driving car/ motorbike, walking, biking bicycle, or riding wheelchair) are summarized. Each of these modes of travel is analyzed against certain factors (all columns in the table except the last one): speed (possible speed in a given mode of travel); route length (total distance and number of segments a user typically travels in a given mode of travel); structure (environment in which a user would travel); and accuracy (level of positional accuracy suitable for a given mode of travel). The last column of the table highlights ambiguous cases; these are situations in each mode of travel that may cause ambiguity during navigation.

Understanding these factors and their relationships to geo-positioning sensors will have benefits to vendors and users. It will assist vendors in selecting one or a combination of appropriate geo-positioning sensors that address the specific needs of applications and it will assist users in realizing performance and cost issues

Table 2.3 Factors in choosing geo-positioning sensors appropriate for outdoor navigation

Mode of Travel	Speed*	Route Length**	Structure	Accuracy	Ambiguous Cases
Driving Car/ Motorbike	Fast	Long	Road segments	Better than 10 m	Intersections Close parallel roads (within geo-positioning accuracy range)
Walking (Pedestrian)	Slow	Short	Sidewalk segments	Better than 1 m	Intersections Narrow roads (close sidewalks on each side of a road)
Biking Bicycle	Medium	Medium	Road/Sidewalk segments	Better than 5 m	Intersections Close parallel roads (within geo-positioning accuracy range) Narrow roads (close sidewalks on each side of a road)
Riding Wheelchair	Slow	Short	Sidewalk segments	Better than 1 m	Intersections Narrow roads (close sidewalks on each side of a road)

*Speed: slow < 10 km/h; 10 km/h < medium < 40 km/h; fast > 40 km/h
**Route Length: short (2-3 short segments); medium (3-6 short/long segments); long (6 or more short/long segments)

associated with navigation systems/services. Each of these four factors is described below.

Speed. Speed at which a user travels is one factor in determining a geo-positioning sensor suitable for outdoor navigation applications. In order to better understand the impact of speed on navigation in outdoors, mobility of a user can be categorized into slow (less than 10 kilometers per hour), medium (between 10 and 40 kilometers per hour), and fast (over 40 kilometers per hour). For example, a geo-positioning sensor with an accuracy range of 10 meters is not suitable for a mode of travel with slow speed. This is because the precise location of a pedestrian with a speed of 1 meter per second (average walking speed) is unknown within 10 meters, even though the user may have changed his/her location several times.

Route length. Route length is another factor that can assist in understanding the scale of space in which a user can travel. Route length can be categorized into short (2-3 short segments), medium (3-6 short/long segments), and long (6 or more short/ long segments). Knowing the number of segments in a route and the length of each segment is important factor affecting the performance of navigation systems/services (number of segments and length of each segment are issues important to map matching algorithms which will be discussed later in this section).

Structure. Structure is a reference to the environment in which the actual physical movements occur. For drivers (cars or motorbikes), structure contains road segments; for pedestrians, structure contains sidewalk segments; for cyclists (bicycles), structure contains both road segments and sidewalk segments; and for wheelchair users, structure contains sidewalk segments. Each of these structures imposes certain constraints on the use of geo-positioning sensors appropriate for navigation in outdoors. For example, GPS-based navigation systems/services are problematic for wheelchair-bound individuals, with sidewalk segments as structure, as sidewalks are often adjacent to buildings that may obstruct GPS signals leading to accuracy degradation (due to multipath problems) or signal blockage.

Accuracy. Accuracy is a reference to the acceptable level of difference between the estimated position and the true position of the user in a given mode of travel. The accuracy ranges in Table 2.3 are not exact as each application in each mode of travel may require a certain finer range. As mentioned earlier, these accuracy levels are partially a function of speed of user which in turn is a function of mode of travel. For driving, which typically involves fast speed, an accuracy range of better than 15 meters is needed. For walking, which typically involves slow speed, an accuracy range of better than 1 meter is needed. For biking, which typically involves medium speed, an accuracy range of better than 5 meters is needed. For riding wheelchairs, which typically involves slow speed, an accuracy range of better than 1 meter is needed.

The last column in Table 2.3 highlights ambiguous cases where navigation systems/services, based on the mode of travel and structure factors, must resolve. For driving, where road segments are used, intersections and close parallel roads (closer than positional accuracy) may cause ambiguities. For walking, where sidewalk segments are used, intersections and narrow roads (i.e., close sidewalks on each side of a road) may cause ambiguities. For biking, where both road segments and sidewalk segments are used, intersections, close parallel roads (closer than positional accuracy), and narrow roads (i.e., close sidewalks on each side of a road) may cause ambiguities. For riding wheelchairs, where sidewalks are used, similar to walking, intersections and narrow roads (i.e., close sidewalks on each side of a road) may cause ambiguities.

Clearly no single geo-positioning sensor can address all the needs of outdoor navigation. For this reason, an alternative has been taking a hybrid approach, combining two or more geo-positioning sensors for those outdoor navigation applications where a single geo-positioning sensor is insufficient. Determination of a combination of geo-positioning sensors that is appropriate for an application depends on the requirements of the underlying application such as accuracy, availability, reliability, and cost. For example, a common combination is GPS with dead reckoning, where a dead reckoning sensor, such as differential odometer, can augment GPS, especially when GPS signals are unavailable.

2.2.2 *Wireless Communication*

While the first and second generations of outdoor navigation technology did not offer any connection to other systems and resources, the third and fourth genera-

Table 2.4 Wireless communication systems for outdoor navigation

Infrastructure	Coverage	Technology	Bandwidth	Cost to User
Global	Anywhere on earth	Satellite communication systems	100 Kb/s	Receiver, service
Wide	City, state, region	Cell based	1 Mb/s	Mobile device, service
Local	Campus, city	WiFi	110 Mb/s	Device (HW/SW), service

tions of outdoor navigation technology is able to connect to external systems and resources through wireless communication systems. The availability of a wireless communication system in an outdoor navigation system/service makes it possible for the system/service to access real-time information, such as traffic, weather, and accidents, that can be used in computing optimal routes. One possible use of wireless communication systems in outdoor navigation services is for computing navigation functions remotely (e.g., on remote servers) and transmitting the results to users (e.g., on smartphones); navigation services have become the emerging trend in providing navigation assistance.

Wireless communication systems that can be employed in outdoor navigation systems/services can be divided into three infrastructure categories (Table 2.4): global, wide, and local. Global infrastructure-based wireless communication systems coverage area spans anywhere on the earth. The best example of the global infrastructure-based wireless communication systems is satellite communication systems featuring 100 Kb/s bandwidth. The cost to the user includes receivers and services.

Wide infrastructure-based wireless communication systems coverage area spans cities, states, and regions. Cell-based communication is the best example of the wide infrastructure-based wireless communication systems featuring 1 Mb/s bandwidth. The cost to the user includes mobile devices and services.

Local infrastructure-based wireless communication systems coverage area spans campuses and cities. WiFi communication is the best example of the local infrastructure-based wireless communication systems featuring 110 Mb/s bandwidth. The cost to the user includes devices (hardware and software) and services.

2.2.3 Database

As data for navigation often is not available in one place, a variety of sources are considered to obtain navigation data. Table 2.5 shows geometrical and topological data, along with other data, that a navigation system/service must contain. For navigation in outdoors, depending on mode of travel, two types of databases are needed: road networks and sidewalk networks. A road network database consists of road segments and a sidewalk network consists of sidewalk segments on both side of a road segment. The essential spatial data in navigation systems/services in-

Table 2.5 Characteristics of data and databases for outdoor navigation

Database	Spatial Data (geometry)	Type	Non-Spatial Data	Navigation Data	Database Size	Functions
Road Segment	Series of coordinates on segments making up segment shape	Geometry, vector	Name, length, width, speed limit, number of lanes, one-two way	POI addresses, landmarks	Very large	Retrieval, mapping
Sidewalk Segment	Series of coordinates on segments making up segment shape	Geometry, vector	Name, length, width, surface type, surface condition	Building addresses, ramps, landmarks	Very large	Retrieval, mapping
Road Network	Coordinates of segment end points	Topology (intersections connectivity), vector	Nodes: intersections Links: road segments	Weight on each road segment (e.g., distance and travel time)	Very large	Routing, direction
Sidewalk Network	Coordinates of segment end points	Topology (intersections connectivity), vector	Nodes: intersections Links: sidewalk segments	Weight on each sidewalk segment (e.g., distance and elevation)	Very large	Routing, direction

cludes road/sidewalk segments where each segment contains a series of coordinates representing its shape (geometry). Road/sidewalk networks in navigation systems/ services primarily facilitate computation of routes and directions. There are two important points worth mentioning here. One is that a network (road or sidewalk) is needed in navigation systems/services as it represents topology (connectivity) between intersections (nodes of the network) and road/sidewalk segments (links of the network). The second point is that network quality, especially accuracy, is of importance for providing accurate and reliable navigation assistance.

Note that in this table, and throughout this chapter, we make a distinction between non-spatial data and navigation data. By non-spatial data we refer to the types of data that describe road/sidewalk networks and segments and by navigation data we refer to the types of data that are specifically designated for navigation purposes.

Figure 2.4 shows the road network and the sidewalk network within the University of Pittsburgh's main campus. Figure 2.4 (a) shows the road network within the campus, Figure 2.4 (b) shows the pedestrian network within the campus, and Figure 2.4 (c) shows the road network and the pedestrian network within the campus overlaid.

In addition to topological information that a network (road or sidewalk) supports, a road network usually should contain such non-spatial data as road segment name, length, width, speed limit, number of lanes, and right-of-way (one- vs. two-way). Similarly a sidewalk network contains such non-spatial data as sidewalk segment name (really road segment name), length, width, surface type, and surface condition. Navigation data typically associated with road segments include POI (e.g., restaurant, gas station, and landmark). Navigation data associated with sidewalk segments include building address, ramp, and landmark.

Navigation data typically associated with road networks include weight on each road segment (e.g., distance, travel time) and navigation data associated with sidewalk networks include weight on each sidewalk segment (e.g., distance, elevation). Routing algorithms utilize navigation data either directly, as the weight, or indirectly through a weight function to derive other weights for computing optimal routes.

Navigation databases that include road/sidewalk networks, segments, non-spatial data, and navigation data are typically very large requiring large capacity storage devices and efficient algorithms for data retrieval.

In general, road/sidewalk segments and associated non-spatial data and navigation data are used for retrieval and mapping functions and road/sidewalk networks and associated non-spatial data and navigation data are used for routing and direction functions.

Figure 2.5 shows an Entity-Relationship (ER) diagram for navigation databases that provide navigation assistance in outdoors with driving as mode of travel. As shown in the figure, road networks constitute the core of these databases. The main entities in this ER diagram are nodes, points, links, and POIs. Nodes, in road networks, are end points of road segments and are identified by Node_ID. A node is a point which has attributes such as Point_ID and coordinates (e.g., latitude and longitude). A series of points (excluding end points) construct the shape of a road segment. A link (road segment) has a variety of attributes such as Link_ID, length,

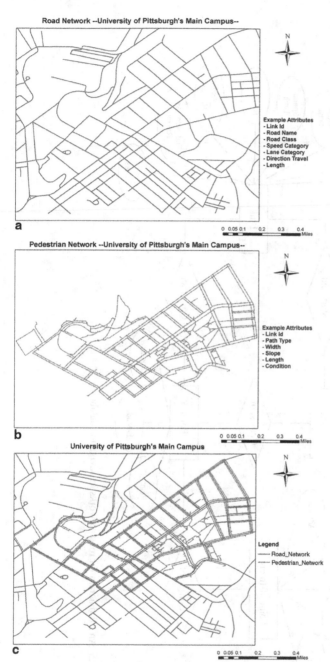

Fig. 2.4 Road and sidewalk networks within the University of Pittsburgh's main campus. **a** Road network within the University of Pittsburgh's main campus. **b** Pedestrian network within the University of Pittsburgh's main campus. **c** Road network and pedestrian network overlaid

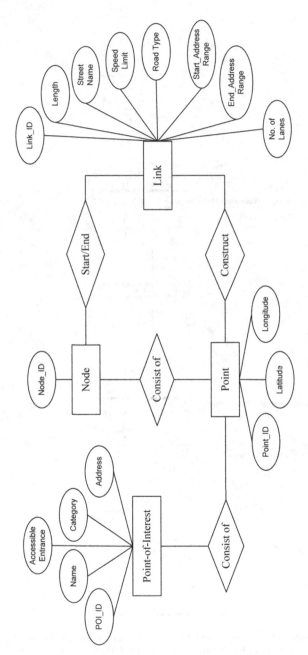

Fig. 2.5 An ER diagram for road networks in navigation databases

street name, path type, slope, surface condition, width, and number of steps. POI is another entity that is a point and has such attributes as POI_ID, name, accessible entrance, category, and address.

Figure 2.6 shows an ER diagram for navigation databases that provide navigation assistance in outdoors with walking as mode of travel. As shown in the figure, sidewalk networks constitute the core of these databases. The main entities in this ER diagram are nodes, points, links, and POIs. Nodes, in sidewalk networks, are end points of sidewalk segments and are identified by such attributes as Node_ID and curb. A node is a point which has attributes such as Point_ID and coordinates (e.g., latitude and longitude). A series of points (excluding end points) construct the shape of a sidewalk segment. A link (sidewalk segment) has a variety of attributes such as Link_ID, length, street name, path type, slope, surface condition, width, and number of steps. POI is another entity that is a point and has such attributes as POI_ID, name, accessible entrance, category, and address.

Road and sidewalk network data are collected, maintained, and disseminated by different providers. Table 2.6 shows the providers of data for road and sidewalk networks which are government agencies or non-profit organizations, commercial mapping companies, and community mapping volunteers. An example of a government agency that provides data for road networks in the United States is Census Bureau; TIGER (Topologically Integrated Geographic Encoding and Referencing system) is one product by the Census Bureau that contains data for navigation purposes. An example of a non-profit organization that provides data for road network data is Pennsylvania Spatial Data Access (PASDA) in Pennsylvania, U.S. Examples of commercial mapping companies that provide data for road networks are NAVTEQ and Tele Atlas. Examples of community mapping volunteers are OpenStreetMap and Wikimapia. The data for sidewalk networks are often provided by local government agencies, such as county and city. Examples of commercial mapping companies that provide data for sidewalk networks in some cities are NAVTEQ and Tele Atlas. An example of a community mapping volunteers is OpenStreetMap.

As for data collection approaches by government agencies and non-profit organizations, satellite imagery, GPS data collection, mobile mapping systems, field survey and paper map digitization and scanning are common. These same approaches are also taken by commercial mapping companies. However, possible data collection approaches by community mapping volunteers are GPS data collection and online manual map digitization.

Advancements in satellite imagery and availability of high-resolution spatial and temporal satellite images have made image processing, to extract data/objects of interest, an attractive and active field of study in the geospatial community. Satellite imagery has become an essential source of data for GIS databases and navigation systems/services. Through automated and semi-automated techniques, data for road and sidewalk network databases can be extracted from satellite images. As shown in Table 2.5, governments, organizations, and commercial companies routinely utilize satellite imagery to collect geospatial data for road and sidewalk network databases. In particular, high-resolution images which provide details at the level appropriate for data in navigation systems/services play a dominant role in collecting and

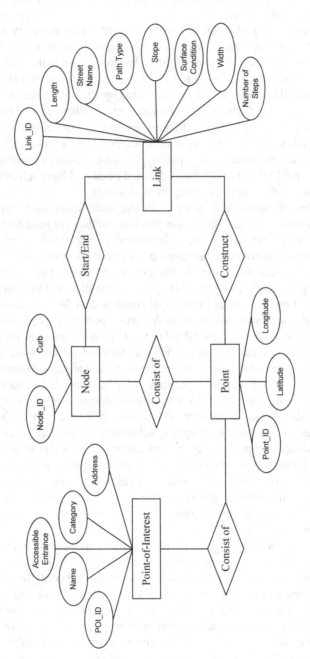

Fig. 2.6 An ER diagram for sidewalk networks in navigation databases

Table 2.6 Map data providers for road and sidewalk networks

Data Providers	Road Network	Sidewalk Network	Data Collection
Government Agencies	E.g., U.S. Census Bureau's TIGER	Local government agencies: county, city	Satellite imagery; GPS data collection;
Non-Profit Organizations	E.g., Pennsylvania Spatial Data Access (PASDA)		Mobile Mapping Systems; Field survey; Paper map digitization and scanning
Commercial Mapping Companies	E.g., NAVTEQ, Tele Atlas	E.g., NAVTEQ (some cities), Tele Atlas (some cities)	Satellite imagery; GPS data collection; Mobile Mapping Systems; Field survey; Paper map digitization and scanning
Community Mapping Volunteers	E.g., OpenStreetMap, Wikimapia	E.g., OpenStreetMap (some areas)	Online GPS data collection; Online manual map digitization

updating road and sidewalk network databases. Figure 2.7 (a)[1] shows an example of a 1 m resolution satellite image and Figure 2.7 (b)[2] shows an example of a 0.3 m resolution satellite image for the same area.

Quality of a road network or a sidewalk network is crucial in the operation of an outdoor navigation system/service as such networks constitute the foundation for many data processings and computations in navigation systems/services. Due to availability of various methods for collecting road data and of various techniques to generate a road network database, different map providers may provide map databases with significantly different qualities. Figure 2.8 shows two examples of road network data, obtained through three different sources (PASDA, TIGER/Line, and NAVTEQ), overlaid and verified with 0.305 m pixel resolution natural color orthoimages obtained from the United States Geological Survey. The accuracy of orthoimages is estimated not to exceed 3 m root mean square error (RMSE). TIGER/Line has the lowest positional accuracy as shown in the figure where some street centerlines intersect buildings. NAVTEQ has a higher resolution and positional accuracy (e.g., for representation of cul-de-sac features) than the other two sources as shown on the right image in Figure 2.8.

Table 2.7 shows the different types and sources of uncertainties in road and sidewalk networks. Three general sources of uncertainties are: geometry, topology, and attribute. Uncertainties associated with geometry in road and sidewalk networks are of three types: accuracy, completeness, and resolution. Uncertainties associated with topology in road and sidewalk networks are of two types: accuracy and com-

---

[1] http://www.pasda.psu.edu/uci/MetadataDisplay.aspx?entry=PASDA&file=doq99_pa.xml&dataset=1

[2] http://www.pasda.psu.edu/uci/MetadataDisplay.aspx?entry=PASDA&file=pittsburghpa2005a_orthoimagery.xml&dataset=603

Fig. 2.7 Example high-resolution satellite images. **a** 1 m resolution. **b** 0.3 m resolution

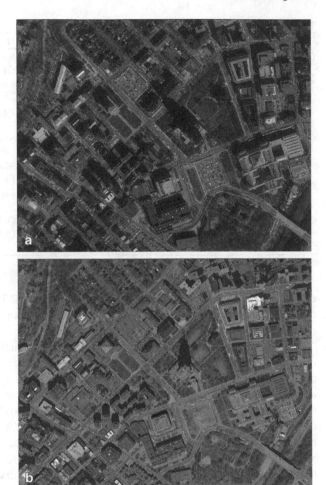

pleteness. Uncertainties associated with attribute in road and sidewalk networks are of two types: accuracy and completeness.

In road networks, examples of uncertainties associated with geometry under accuracy type include inaccurate coordinates of nodes (e.g., the represented location of an intersection is very far from its true location) and inaccurate coordinates of points representing links (e.g., mismatch between the points forming a road segment and the true location of the road segment); examples of uncertainties associated with geometry under completeness type include missing nodes (e.g., an intersection is not stored in the database) and missing links (e.g., a road segment is not stored in the database); an example of uncertainties associated with geometry under resolution type is insufficient points representing segments (e.g., the small number of points on a road segment do not represent the true shape of the road segment).

In road networks, an example of uncertainties associated with topology under accuracy type is incorrect connectivity at junctures (e.g., an intersection does not have

Fig. 2.8 Digital road network data obtained through three different sources

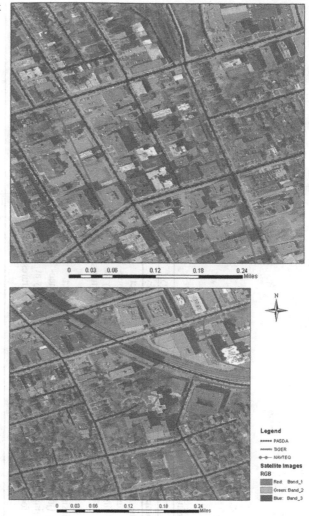

the correct connection to road segments); and an example of uncertainties associated with topology under completeness type is missing nodes (e.g., an intersection is not stored in the database).

In road networks, examples of uncertainties associated with attribute under accuracy type are inaccurate name of a road segment, type of a road segment, and number of lanes in a road segment; examples of uncertainties associated with attribute under completeness type are missing road segment, road segment type, and number of lanes in a road segment.

In sidewalk networks, an example of uncertainties associated with geometry under accuracy type include inaccurate coordinates of nodes (e.g., a decision point is located in a wrong place) and inaccurate coordinates of points representing links

Table 2.7 Types and sources of uncertainties in road and sidewalk networks

Network	Geometry			Topology		Attribute	
	Accuracy	Completeness	Resolution	Accuracy	Completeness	Accuracy	Completeness
Road	Inaccurate coordinates of nodes (junctures); inaccurate coordinates of points representing links (segments)	Missing nodes (junctures); Missing links (segments)	Insufficient points representing segments	Incorrect connectivity at junctures	Missing nodes (junctures)	Inaccurate name, type, number of lanes	Missing name, type, number of lanes
Sidewalk	Inaccurate coordinates of nodes (junctures); inaccurate coordinates of points representing links (segments)	Missing nodes (junctures); missing links (e.g., footpath)	Insufficient points representing segments	Incorrect connectivity at junctures	Missing nodes (junctures)	Inaccurate name, side, type	Missing name, side, type

(e.g., the points representing a sidewalk segment do not match the true location of the sidewalk segment); examples of uncertainties associated with geometry under completeness type are missing nodes (e.g., a decision point is not stored in the database) and missing links (e.g., a footpath is not included in the database); an example of uncertainties associated with geometry under resolution type is insufficient points representing segments (e.g., the small number of points on a footpath do not represent the true shape of the footpath).

In sidewalk networks, an example of uncertainties associated with topology under accuracy type is incorrect connectivity at junctures (e.g., a decision point does not have the correct connection to sidewalk segments); an example of uncertainties associated with topology under completeness type is missing nodes (e.g., a decision point is not stored in the database).

In sidewalk networks, examples of uncertainties associated with attribute under accuracy are inaccurate name of sidewalk segment, side of the road to which sidewalk segment belongs and type of sidewalk segment; examples of uncertainties associated with attribute under completeness are missing name of sidewalk segment, side of the road to which sidewalk segment belongs and type of sidewalk segment.

2.3   Functions

An outdoor navigation system/service performs a variety of functions, some are obvious to users and some are performed in the background. Table 2.8 shows the main functions performed by most outdoor navigation systems/services.

Retrieval. The retrieval function is responsible for retrieval of data from the database which contains spatial and attribute data. The input to the retrieval function is usually a POI name and the output, depending on the input, could be a location of an intersection or a POI address.

Map creation. The map creation function is responsible for creating a map using the centroid of an area (e.g., city), a POI location, or current location (obtained through geo-positioning sensors). This function needs all data including road/sidewalk networks, road/sidewalk segments, and attribute data.

Mapping. Once a map is created, the user is able to perform mapping functions. The mapping function allows user to zoom in, zoom out, and pan the created map. Similar to the map creation function, the input to the mapping function could be the centroid of an area (e.g., city), a POI location, or current location and the output is a new map. All available data including road/sidewalk networks, road/sidewalk segments, and attribute data are needed for this function.

Geocoding. The geocoding function is responsible for computing the coordinates of an address or a POI. The input to the geocoding function often is a POI address and the output is the location of the POI address on the map. The geocoding process in most cases involves an interpolation scheme whereby spatial information about the end nodes of road/sidewalk segments (i.e., coordinates of end points), the geometry of the segment (i.e., series of coordinates forming the shape of the

Table 2.8 Main functions performed by outdoor navigation systems/services

Computation	Input	Process	Output	Database
Retrieval	POI name	Retrieve spatial, non-spatial data	Location of intersection, address of POI	Spatial, attribute
Map Creation	Centroid location, POI location, current location	Clip spatial and non-spatial data	Map	Road/sidewalk segments, attribute
Mapping	Centroid location, POI location, current location	Zoom in, zoom out, pan	Map	Road/sidewalk segments, attribute
Geocoding	Address	Interpolation	Location on map	Road/sidewalk segments, attribute
Routing/Rerouting	Origin-Destination addresses (current location for rerouting)	Optimal routing	Route on map	Road/sidewalk networks
Tracking	Position data	Map match position data on road/sidewalk segments	Current location on map	Road/sidewalk networks, road/sidewalk segments, attribute
Direction	Route	Compute distance and search for landmarks	A set of instructions to navigate from origin (or current location) to destination	Road/sidewalk networks, road/sidewalk segments, attribute

segment), and the address ranges on both sides of the segment (i.e., start and end addresses on each side of the segment) are used to estimate the location of a given address. The data needed for the geocoding function includes POI address, road/ sidewalk segments, and attribute data.

Geocoding is a common function in outdoor navigation systems/services as well as other applications and services such as Web Mapping Services (WMSs), e.g., Google Maps, that provide navigation assistance. Results of geocoding by different WMSs are illustrated in Figure 2.9. Figure 2.9 (a) shows a geocoded location from Google's street geocoding service where the geocoded point is on the side of the building's main entrance. Figure 2.9 (b) shows multiple locations of an address geocoded through different geocoding services.

A reference database upon which a geocoding algorithm geocode addresses, through interpolation, plays an important role in the process. Street centerlines con-

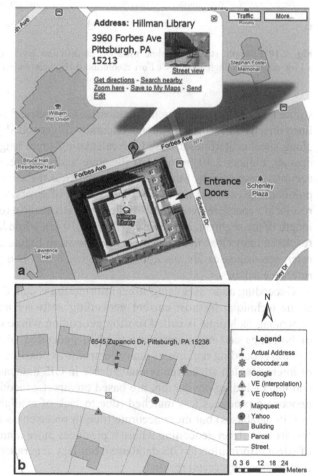

Fig. 2.9 Example geocoding results by different services. **a** Geocoding by Google. **b** Geocoding by multiple services

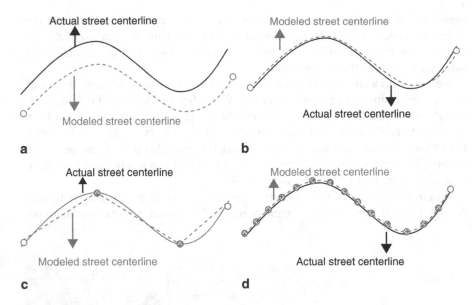

Fig. 2.10 Example street centerline modeled at different levels of accuracy and resolution.
a Low accuracy **b** High accuracy **c** Low resolution **d** High resolution

stitute the core of such reference databases. Geometrical accuracy of street centerlines directly influences positional accuracy of geocoded points since coordinates
along street centerlines are used for calculating street length and interpolating addresses. Figures 2.10 (a) and (b) illustrate a street centerline with low and high
accuracy, respectively. Street resolution, which is the sampling frequency of shape
points along each segment, is another source of errors. Low-resolution sampling
results in a coarse representation of the actual street and a rough estimate of the total
length of the segment, as shown in Figure 2.10 (c), while high-resolution sampling
results in lines close to actual streets, as shown in Figure 2.10 (d). Low-resolution
reference databases are susceptible to higher errors compared to high-resolution
reference databases.

Geocoding addresses using street information is called street geocoding, which
is the technique in most current geocoding software and tools. An alternative
geocoding technique is called rooftop geocoding where coordinates of centroids
of buildings and monuments are used to geocoded addresses. However, since
rooftop geocoding requires as many centroids as possible in the geographic extent
of interest which are currently unavailable in GIS databases, most existing applications, including navigation, are based on street geocoding. In general, rooftop
geocoding produces less matched (due to lack of available centroid coordinates
in GIS databases) but more accurate (due to preciseness of centroid coordinates)
results than street geocoding where it produces more matched (due to availability
of street information in GIS databases) but less accurate (due to impreciseness of
information on streets) results.

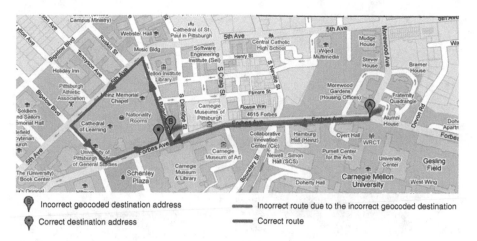

Incorrect geocoded destination address

Incorrect route due to the incorrect geocoded destination

Correct destination address

Correct route

Fig. 2.11 Two routes from single origin to two different locations

Destinations in navigation systems/services can be obtained through on-the-fly geocoding or previously geocoded and stored data, thus susceptible to errors due to uncertainties in geocoding process and map database. Incorrect geocoding, on-the-fly or stored, results in wrong locations of addresses as destinations and undesired routes. Figure 2.11 shows two routes, one from origin A to the correct location of a destination, and another from origin A to an incorrect location of the destination estimated by geocoding.

Routing/Rerouting. The routing function is responsible for computing optimal routes, based on a pre-determined criterion, between pairs of origin-destination addresses. The origin could be entered by the user or current user's location obtained through geo-positioning sensors. The output is the computed route highlighted on the map. The main data needed for the routing function is the road/sidewalk network within the geographic extent of interest (this can be determined by the area that covers both origin and destination locations) which provides the topology of the network, a requirement by all routing algorithms. Today's navigation systems/ services allow different routing criteria such as shortest distance, fastest travel time, least intersections, and no tolls among others. An extension to the routing function is rerouting which, depending on the situation (e.g., deviation from the computed route), re-computes a new route from user's current location, as the origin, to the destination.

In general, routes in navigation systems/services could be computed through exact algorithms or heuristic algorithms. Exact algorithms are those algorithms that consider all possible options between pairs of origin and destination addresses to find optimal solutions. Heuristic algorithms are those algorithms that are based on rule-of-thumbs (shortcuts) to find good solutions between pairs of origin and destination addresses. In other words, exact algorithms consider the entire solution space (i.e., all possible routes) to find optimal solutions and heuristic algorithms consider a portion of the solution space (i.e., a subset of all possible routes), to find solutions

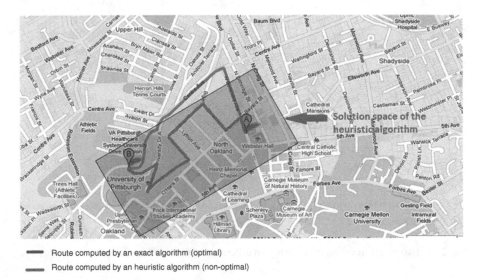

■ Route computed by an exact algorithm (optimal)
▬ Route computed by an heuristic algorithm (non-optimal)

Fig. 2.12 An optimal route computed by an exact algorithm and a non-optimal route computed by a heuristic algorithm

which may not necessarily be optimal. Factors affecting choice of exact or heuristic algorithms for navigation systems/services include acceptable response time, networks size, and computational power. Acceptable response time is an important factor in navigation systems/services for computing routes, especially rerouting that is typically computed in real time while the vehicle is moving. Network size is determined by the total number of nodes (e.g., intersections) in a network. In general, the larger the number of nodes, the longer the computation will be. Computational power of a navigation device is another factor that impacts performance of routing and rerouting. Figure 2.12 illustrates an example of an optimal route (in blue) computed by an exact algorithm and an example of a good (non-optimal) route (in red) computed by a heuristic algorithm. In this example, the strategy by the heuristic algorithm is to trim the entire network (solution space) to a smaller network (a window around the origin and destination locations as shown in Figure 2.12). By using the sub-network (the network within the window), the heuristic algorithm only considers some of the routes surrounding the origin A and the destination B locations to find a solution, which may be acceptable but not optimal.

Other than incorrect geocoded destination addresses, which are of geometrical errors, topological errors and attribute errors in networks result in incorrect routes as well. Figure 2.13 shows an example where the computed shortest route (in blue) is different and longer, due to topological errors (e.g., a road segment is incorrectly stored as dead-end), than the actual shortest route (in red). Figure 2.14 shows an example where attribute errors (e.g., incorrect road segment orientation, one- or two-way) result in a route which is different and longer than the actual shortest route.

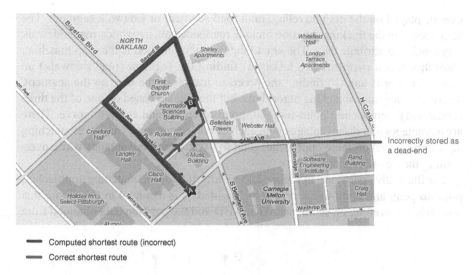

━━━ Computed shortest route (incorrect)

━━━ Correct shortest route

Fig. 2.13 An example of an incorrect route due to topological errors

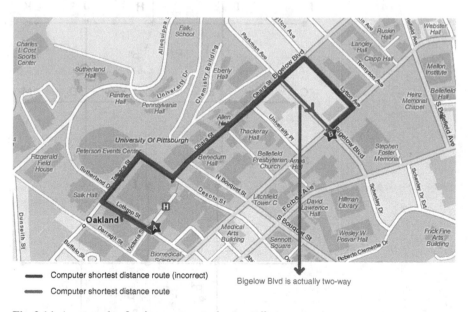

━━━ Computer shortest distance route (incorrect)   Bigelow Blvd is actually two-way

━━━ Computer shortest distance route

Fig. 2.14 An example of an incorrect route due to attribute errors

Tracking. The tracking function is responsible for the continuous monitoring of user's location in real time. The input to the tracking function is the position data acquired continuously from geo-positioning sensor(s) at a fixed interval (time or distance) where it is used for map matching. The output is the real-time location of the user displayed on the map traveling, on a road segment or sidewalk segment. The data needed in the tracking function include road/sidewalk networks, road/sidewalk segments, and attribute data. The key to the tracking function is the map matching algorithm which performs two tasks: (a) finding the segment (road/sidewalk) on which the user is and (b) finding the precise location of the user on the segment. There are many map matching algorithms, but they all are based on one of the three fundamental approaches: point-to-point, point-to-curve, and curve-to-curve. There are advantages and disadvantages with each of these approaches and map matching algorithms based on these approaches tend to exploit their advantages while overcoming their disadvantages.

For illustrative purposes, in Figure 2.15 results of two map matching approaches, point-to-point and point-to-curve, are highlighted. It is assumed in this figure that a vehicle is travelling on segments AB and BD and the position data obtained from

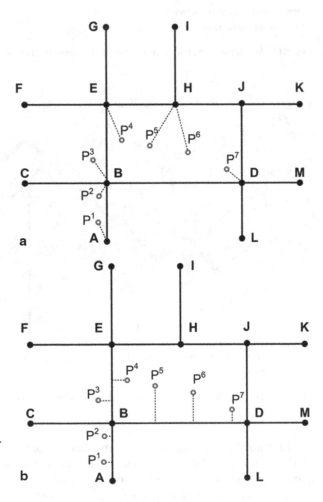

Fig. 2.15 Examples of map-matched results using different algorithms. **a** Point-to-point map matching. **b** Point-to-curve map matching

Fig. 2.16 Example direction generated between a pair of addresses

A 135 N Bellefield Ave
   Pittsburgh, PA 15213

1. Head **south** on **N Bellefield Ave** toward **5th**   374 ft
   Ave

2. Take the 1st **right** onto **5th Ave**   0.5 mi

3. Turn **right** at **Lothrop St**   295 ft
   Destination will be on the right

B 190 Lothrop St
   Pittsburgh, PA 15213

geo-positioning sensors are indicated by P^1 to P^7. The map-matched results of the point-to-point approach in Figure 2.15 (a) show that the vehicle travels from AB to BE to EH and to BD. The map-matched results of the point-to-curve approach in Figure 2.15 (b) show that the vehicle travels from AB to BE and to BD. Both algorithms produce incorrect results for P^3 and P^4.

Direction. The direction function uses the computed route to provide instructions on how to travel from the origin to each segment of the route to reach the destination. The input to the direction function is a route and the output is a set of instructions to navigate on the route displayed on the map and/or presented through voice. The set of instructions basically utilizes information on intersections, distances on road segments, and landmarks. The input to the direction function includes road/sidewalk networks, road/sidewalk segments, and attribute data.

Direction generation refers to the process of generating step-by-step instructions for a given route, which involves two main steps. First, instructions are generated for the entire route from origin to destination. This information is usually delivered to the user as text. Figure 2.16 shows an example of such directions. Second, the instructions generated in the first step are presented to the user turn-by-turn in real time, using the tracking function, based on vehicle's position.

In general, navigation, and consequently directions, could be simple or complex. Factors impacting directions to be simple or complex include navigation environment (structure and density), route length, route complexity in terms of decision points, and user's familiarity with the navigation environment. Figures 2.17 (a) and (b) show examples of routes and directions on them. In Figure 2.17 (a) direction is simple as only one road segment exists in the computed route, whereas in Figure 2.17 (b) direction is complex as there are many decision points where the user must make decision.

Each step of direction generation introduces some uncertainties, mainly associated with road attributes, to the resultant directions. The main sources of errors in the first step of direction generation are segment name and segment length. The sources of errors in the second step of direction generation are associated with vehicle's position determination at each time epoch using map matching and distance

a

Road A

10km

1. Go straight forward on road A for 10km

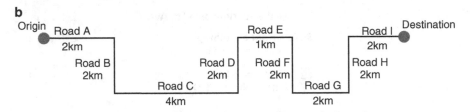

b

Origin Road A   Road E   Road I Destination
        2km   1km   2km
    Road B   Road D Road F   Road H
    2km   2km   2km   2km
        Road C   Road G
        4km   2km

1. Go straight forward on road A for 2 km   10. Turn right on road F
2. Turn right on road B   11. Go straight forward on road F for 2 km
3. Go straight forward on road B for 2 km   12. Turn left on road G
4. Turn left on road C   13. Go straight forward on road G for 2 km
5. Go straight forward on road C for 4 km   14. Turn left on road H
6. Turn left on road D   15. Go straight forward on road H for 2 km
7. Go straight forward on road D for 2 km   16. Turn right on road I
8. Turn right on road E   17. go straight forward on road I for 2 km
9. Go straight forward on road E for 1 km

Fig. 2.17 Example directions. **a** Simple directions. **b** Complex directions

Fig. 2.18 An example of direction errors due to incorrect street name

Driving directions to 300 Waterfront Dr W, West Homestead, PA 15120
1.6 mi – about 4 mins

Ⓐ 4250 Murray Ave
   Pittsburgh, PA 15217

1. Head **south** on **Murray Ave** toward **Loretta St**   0.2 mi

2. Turn **left** at Hazelwood Ave ←   ─235 ft─ **Forward Ave (wrong street name)**

3. Take the **1st right** onto **Browns Hill Rd**   0.6 mi

4. Continue onto **Homestead Grays**   0.5 mi

5. Turn **right** at **5th Ave**   0.2 mi

6. Slight **right** at **Waterfront Dr W**   0.1 mi
   Destination will be on the right

Ⓑ 300 Waterfront Dr W
   West Homestead, PA 15120

Save to My Maps
Sponsored Links

estimation to the next decision point based on vehicle's position, speed, and current road segment length. Figure 2.18 shows an example of generated directions with an incorrect street name.

2.4   Static and Dynamic Navigation

Outdoor navigation systems/services can operate either in static mode or in dynamic mode. In static mode (or stand-alone mode), the navigation system/service uses only the data stored in its database and it has no access to real-time information such as traffic, weather, or accidents. Rerouting in static mode occurs only when the user deviates from the computed route, in which case the new location of the user is used to compute a new route to the destination.

In dynamic mode, in addition to the stored database the navigation system/service has access, through a wireless communication system, to remote resources. Such remote resources could be real-time weather information, real-time traffic updates, real-time road accidents, among others, within the traveling range. Rerouting in dynamic mode may occur when the user deviates from the computed route or when new real-time information affecting the travelling route is made available to the system/service. In this case, the navigation system/service uses the current location of the user to compute a new route (which may contain a portion of the original route computed at the start of the trip) by taking into account the newly acquired real-time information. For example, if user's preference for routes is least congested, then the navigation system/service capable of dynamic mode will continually check for traffic information on the upcoming segments of the original computed route. If the information it receives indicates congestion on the segments that the user is about to reach, it re-computes a new route to the destination, replaces it with the original computed route and presents it to the user.

2.5   User

The technologies and techniques that are appropriate for outdoor navigation were discussed in the previous sections. In this section, we focus on usability, another important aspect of navigation systems/services. Usability in outdoor navigation systems/services is defined as the accessibility of navigation functions and features by its users. Given that outdoor navigation systems/services can benefit from a wide range of hardware and technologies and that the demand for navigation systems/services is continually increasing, navigation vendors are paying more attention to the usability aspect of their products. For an outdoor navigation system/service to be able to provide appropriate guidance, it is required that it supports various features such as map representation, navigation situations, purpose of trip, and user preferences. Table 2.9 shows the usability features for different modes of travel in outdoor navigation systems/services.

In Table 2.9 users are categorized based on mode of travel, (i.e., car/motor driver, pedestrian, bike rider, and wheelchair rider). For each group, map presentation, navigation situation (static or dynamic), purpose of trip, and user preferences with respect to POI, route, and map presentation are analyzed and described below.

Table 2.9 Usability features in outdoor navigation systems/services

Mode of Travel	Map Presentation	Navigation Situation		Purpose of Trip	User Preferences		Map Presentation
		Static	Dynamic		Route	POI	
Driver/Motor rider	Streets; Roads; Landmarks	Safety; Toll	Traffic; Weather; Time; Accident; Route under construction	Commute	Shortest path; Fastest time; Least intersections; Avoid tolls; Least left turns; Safe route; Low congestion	Office; School; Home; Library	Voice/text directions; Color; Brightness; Screen size; Font size; Map scale; Selected POIs
				Emergency	Shortest path; Fastest time; Low congestion	Hospital; Clinic; Pharmacy	
				Leisure	Scenic route; Avoid tolls; Safe route; Low congestion	Restaurant; Bar; Shopping mall	
Pedestrian	Sidewalk; Buildings; Landmarks	Safety	Weather; Time; Route under construction	Commute	Shortest path; Safe route	Office; School; Home; Library	Voice/text directions; Color; Brightness; Screen size; Font size; Map scale; Selected POIs; Unsafe neighborhood
				Leisure	Safe route; Scenic route	Shopping center; Restaurant; Park; Bookstore	
Bike Rider	Streets; Roads; Sidewalks; Buildings; Landmarks	Safety; Toll	Weather; Traffic; Time; Accident; Route under constructions	Commute	Shortest path; Least intersections; Least left turns; Route with feasible slope; Safe route	Office; School; Home; Library	Voice/text directions; Color; Brightness; Screen size; Font size; Map scale; Selected POIs; Unsafe neighborhoods
				Leisure	Least intersections; Least left turns; Route with feasible slope; Safe route; Scenic route	Shopping center; Restaurant; Park; Book store	
Wheelchair Rider	Sidewalk; Buildings; Landmarks; Inaccessible routes	Safety; Slopes; Curbs; Steps	Traffic; Weather; Safety; Obstacles	Commute	Shortest path; Least left turns; Route with feasible slope; Safe route; Avoid obstacles; Avoid curbs; Avoid steps	Office; School; Home; Library	Voice/text directions; Color; Brightness; Screen size; Font size; Map scale; Selected POIs; Unsafe neighborhoods; Inaccessible routes

Map presentation: Maps play a major role in an outdoor navigation system/ service as they are the media with which users interact the system/service. A map for navigation purposes must provide the right amount of information, as both in-complete and excessive information may confuse the user; present the orientation dynamically in the direction of travel; and display each set of different objects in a different color. These parameters may be computed and presented differently for each mode of travel. For example, for car/motor drivers, road networks along with relevant objects (e.g., landmarks) on or around road segments that assist in naviga-tion are appropriate. For pedestrians, sidewalk networks and relevant objects (e.g., buildings and landmarks) on or around sidewalk segments are appropriate. For bike riders, since the user may ride on both roads and sidewalks, the presentation may include both road networks and relevant objects on road segments and sidewalk networks and relevant objects on sidewalk segments. In areas where bike riders are required to bike only on designated paths, the presentation must include only such paths and relevant objects on them. For wheelchair riders, sidewalks, build-ings, landmarks, and accessible routes must be presented. Due to the impediments in passing inaccessible routes, such as curbs, steps, obstacles, and slopes with more than a specific amount, by wheelchair riders, the map either must not present them at all or present them in such a way that are not confusing to wheelchair riders.

Building a navigation system/services that is capable of representing maps ap-propriate for each mode of travel as discussed above is complex, which is why most current navigation systems/services are designed for one mode of travel, though new navigation systems/services that are multi-modal are emerging.

Navigation situation: One way to analyze navigation situations for each mode of travel is categorizing them into one of two groups: static and dynamic. Static navigation situations are situations which affect user preferences and do not change over time. Dynamic navigation situations are those which are temporary (i.e., changing over time). For drivers and motor riders, safety and tolls are considered as static situations, whereas weather, accidents, roads under construction, and road side development are considered as dynamic situations. For pedestrians, safety is considered as static navigation situation, and weather, time of day, sidewalks under construction are considered as dynamic situations. For bike riders, since they can ride on both roads and sidewalks, navigation situations are the combination of those for drivers/motor riders and pedestrians. For wheelchair riders, safety, slopes, curbs, and steps are considered as static navigation situations, and traffic, weather, safety, and obstacles are considered as dynamic situations.

Purpose of trip: User's preferences usually vary based on purpose of trip. One may prefer to take a different route while commuting than a route between the same origin and destination locations when purpose of trip is leisure. For instance, con-sider a restaurant next to a user's office. The user drives from his/her house to the office everyday, a specific time, which may be during rush hours, impacting his/her preferences with respect to fastest, shortest or least congested route. However, when going to the restaurant next to his/her office during the weekend, his/her preference might be a scenic route which could be even a longer route.

User's preferences can be categorized (see Table 2.6) based on three different purposes of trip for drivers/motor riders (i.e., commute, emergency, and leisure), two purposes of trip for pedestrian and bike riders (i.e., commute and leisure), and one purpose of trip for wheelchair riders (i.e., commute). For each purpose of trip, user's preferences for routes and POIs are identified. For drivers and motor riders there are several preferences such as shortest, fastest, least turns, non-toll, least left turns, safest, or least congested route when commute is purpose of trip. Typical everyday POIs include office, school, home, and library. In emergency situations, since time is of the essence, shortest, fastest, and least congestion routes are usually preferred. Typical emergency POIs include hospital, clinic, and pharmacy. In addition, preferences for trips chosen as leisure are determined as scenic, non-toll, safest, and least-congested routes. Typical leisure POIs include restaurant, bar, and shopping mall.

Likewise, as Table 2.6 shows, preferences for pedestrians, bike riders, and wheelchair riders are determined based on purpose of trip. However, these preferences are not confined to those aforementioned, and also may vary user by user. The complexity of an outdoor navigation system/service is affected by different needs and preferences they address.

User's preferences presented on maps are usually independent of purpose of trip. These preferences are roughly the same for each mode of travel with minor differences. Regardless of mode of travel and purpose of trip, each user has some preferences for receiving directions (via voice or text) and presenting color and brightness, font size, map scale, and screen size of the device. However, for presenting upcoming POIs and landmarks, users often prefer to be notified of the ones which are related to their purpose of trip (e.g., parks in leisure and hospitals in emergency situations). Moreover, for pedestrians, bike riders, and wheelchair riders presenting unsafe neighborhoods and inaccessible routes for wheelchair riders would be useful.

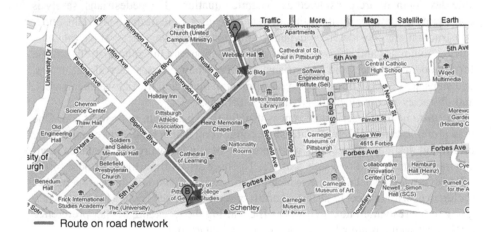

▬▬▬ Route on road network

Fig. 2.19 Computed route on a road network

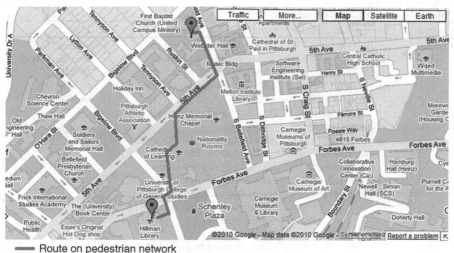

—— Route on pedestrian network

Fig. 2.20 Computed route on a sidewalk network

Figure 2.19 shows an example of a route for outdoor navigation. In this figure the computed route between A and B locations are highlighted on a road network of the navigation environment.

Figure 2.20 shows an example of a route for outdoor navigation. In this figure the computed route between A and B locations are highlighted on a sidewalk network of the navigation environment.

Table 2.10 shows a sample of navigation systems/services currently in the market. Navigation vendors featured in this table are Garmin, TomTom, Magellan, and Pioneer. In this table, the products by each navigation vendor along with their general characteristics are highlighted.

2.6　Summary

In this chapter, characteristics, technologies, and techniques of outdoor navigation systems/services are discussed. The main technologies in navigation systems/services are geo-positioning, wireless communication, and database. Of the possible geo-positioning sensors, GNSS, currently GPS, is the main geo-positioning sensor for outdoor navigation. GPS and its uncertainty are described. Wireless communication systems and their use in outdoor navigation systems/services are discussed. Road and sidewalk networks, composed of road/sidewalk segments, constitute the core of data in outdoor navigation systems/services. The main functions performed by today's modern outdoor navigation systems/services are POI retrieval, map creation, geocoding, routing/rerouting, and tracking. Static and dynamic navigation, users, and routing options in outdoor navigation systems/services are described.

Table 2.10 Sample navigation services

Company	Product Name/Model	General Characteristics	URL
Garmin	nüvi® 3790T *Versions:* nüvi® 3790T nuvi 3790T, Europe, Premium traffic nuvi 3790T, Europe, Premium traffic nuvi 3790T, Austraila and New Zealand, Premium traffic nuvi 3790T, Russia nuvi 3790, Arabic	Map coverage: North America; Europe; Australia; New Zealand; Middle East 4.3" diagonal multi-touch glass display Dual orientation (horizontal, vertical) nüRoute technology with traffic-Trends& MyTrend 3D Building & terrain view Lane assist with junction view Hands-free calling Subscription-free traffic alert Voice command Navigate city transit Calculate more fuel-efficient route Track fuel usage Measurement & currency conversions Emergency locator Anti-theft Nearly 6 million POIs	https://buy.garmin.com/shop/shop.do?cID=134&pID=63940
TomTom	GO 950 LIVE	Map coverage: Western & central Europe, USA, Canada Hands-free calling Enhanced positioning technology (uninterrupted navigation even in tunnels, etc.) IQ Routes technology Advanced lane guidance Local search with Google Live snapshots QuickGPSfix Speed Cameras Weather condition and forecast 4.3" touchscreen Map Share technology to correct maps & benefits from other users' corrections Provide real-time traffic Compute fuel-efficient routes Provide latest fuel price Emergency menu	http://www.tomtom.com/en_gb/products/car-navigation/go-950-live/index.jsp

Table 2.10 (continued)

Company	Product Name/Model	General Characteristics	URL
Magellan	RoadMate 3065	Map coverage: USA, Canada, Puerto Rico 4.7" touchscreen Hands-free calling Life-time free traffic Highway lane assist Built-in AAA TourBook Multi-destination routing Include millions of POIs Automatic night view	http://www.magellangps.com/products/product.asp?segID=354&prodID=2327
Pioneer	AVIC-Z120BT	In-car navigation system Map database: TeleAtlas Map coverage: US, Canada, & Puerto Rico 7" touchscreen Inputing destination addresses by voice Including more than 12 million POIs Estimate fuel cost of trip Estimate vehicle's CO_2 emission Log and archive driving data and analyze driving habits for generating different reports and suggestions for improving fuel efficiency	http://www.pioneerelectronics.com/PUSA/Products/CarAudioVideo/In-Dash/GPS-Navigation-Systems/AVIC-Z120BT

Further Readings

Abdel-Hafez MF (2010) The autocovariance least-squares technique for GPS measurement noise estimation. IEEE Transactions on Vehicular Technology 59:574-588.

Abdel-Hafez MF (2010) On the GPS/IMU sensors' noise estimation for enhanced navigation integrity. Mathematics and Computers in Simulation.

Ali J, Ullah Baig Mirza MR (2010) Performance comparison among some nonlinear filters for a low cost SINS/GPS integrated solution. Nonlinear Dynamics 1-12.

Allain DJ, Mitchell CN (2010) Comparison of 4D tomographic mapping versus thin-shell approximation for ionospheric delay corrections for single-frequency GPS receivers over North America. GPS Solutions 14:279-291.

Alnaqbi A, El-Rabbany A (2010) Precise GPS positioning with low-cost single-frequency system in multipath environment. Journal of Navigation 63:301-312.

Andersson E, Supej M, Sandbakk O, Sperlich B, Stöggl T, Holmberg HC (2010) Analysis of sprint cross-country skiing using a differential global navigation satellite system. European Journal of Applied Physiology 1-11.

Arias J, Lázaro J, Zuloaga A, Jiménez J, Astarloa A (2007) GPS-less location algorithm for wireless sensor networks. Computer Communications 30:2904-2916.

Bączyk R, Kasiński A (2010) Visual simultaneous localisation and map-building supported by structured landmarks. International Journal of Applied Mathematics and Computer Science 20:281-293.

Bai L, Wang Y (2010) A sensor fusion framework using multiple particle filters for video-based navigation. IEEE Transactions on Intelligent Transportation Systems 11:348-358.

Bai L, Wang Y, Fairhurst M (2010) Multiple Condensation filters for road detection and tracking. Pattern Analysis and Applications 1-12.

Balaebail RA, Ariyur KB (2010) Motion estimation and navigational drift correction with LiDAR data. In: NATO/SPS International Technical Meeting on Air Pollution Modelling and its Application, vol. 1, pp 337-345 Torino, Italy.

Barabanova LP (2010) Minimization of GNSS geometric factors. Journal of Computer and Systems Sciences International 49:310-317.

Batista P, Silvestre C, Oliveira P (2010) Optimal position and velocity navigation filters for autonomous vehicles. Automatica 46:767-774.

Bétaille D, Toledo-Moreo R (2010) Creating Enhanced Maps for Lane-Level Vehicle Navigation. IEEE Transactions on Intelligent Transportation Systems.

Cao W, Hauschild A, Steigenberger P, Langley RB, Urquhart L, Santos M, Montenbruck O (2010) Performance evaluation of integrated GPS/GIOVE precise point positioning. vol. 2, pp 710-722.

Cappelle C, Najjar MEBE, Pomorski D, Charpillet F (2010) Intelligent geolocalization in urban areas using global positioning systems, three-dimensional geographic information systems, and vision. Journal of Intelligent Transportation Systems: Technology, Planning, and Operations 14:3-12.

Cesetti A, Frontoni E, Mancini A, Zingaretti P, Longhi S (2010) A Vision-based guidance system for UAV navigation and safe landing using natural landmarks. Journal of Intelligent and Robotic Systems: Theory and Applications 57:233-257.

Chang CL (2010) Using fuzzy logic controller with adaptive detection scheme for fast acquisition of satellite navigation signals. Journal of the Chinese Institute of Engineers, Transactions of the Chinese Institute of Engineers,Series A/Chung-kuo Kung Ch'eng Hsuch K'an 33:367-378.

Chen YH, Juang JC, Kao TL (2010) Robust GNSS signal tracking against scintillation effects: A particle filter based software receiver approach. vol. 2, pp 797-805.

Cheng PC, Lee KC, Gerla M, Härri J (2010) GeoDTN+Nav: Geographic DTN routing with navigator prediction for urban vehicular environments. Mobile Networks and Applications 15:61-82.

Chien YH (2010) Effects of three hypermedia topologies on users' Navigational performance. Asian Journal of Information Technology 9:72-77.

Choi KH, Park SY, Kim SH, Lee KS, Park JH, Cho SI (2010) Methods to detect road features for video-based in-vehicle navigation systems. Journal of Intelligent Transportation Systems: Technology, Planning, and Operations 14:13-26.

Dilellio J (2010) A hybrid GNSS integrity design leveraging a priori signal noise characteristics. Journal of Navigation 63:513-526.

Duckham M, Winter S, Robinson M (2010) Including landmarks in routing instructions. Journal of Location Based Services 4:28-52.

Duric Z, Goldenberg R, Rivlin E, Rosenfeld A (2002) Estimating relative vehicle motions in traffic scenes. Pattern Recognition 35:1339-1353.

Dutta G, Mullur H, Roh JC, Rao S, Hosur S (2008) Improving user experience in assisted land navigation using sub-optimal sensor configuration installed on a cell phone. In: 21st International Technical Meeting of the Satellite Division of the Institute of Navigation (ION GNSS 2008), vol. 4, pp 1901-1906.

Fang J, Gong X (2010) Predictive iterated kalman filter for INS/GPS integration and its application to SAR motion compensation. IEEE Transactions on Instrumentation and Measurement 59:909-915.

Fantino M, Molino A, Nicola M (2010) N-Gene: A complete GPS and Galileo software suite for precise navigation. In: ITM - NATO/SPS International Technical Meeting on Air Pollution Modelling and its Application, vol. 2, pp 1245-1251 Torino, Italy.

Fröhlich P, Schatz R, Leitner P, Baldauf M, Mantler S (2010) Augmenting the driver's view with realtime safety-related information.

Geng T, Zhao Q, Liu J, Li G (2010) Regional orbit determination of navigation satellite based on global priori information. Wuhan Daxue Xuebao (Xinxi Kexue Ban)/ Geomatics and Information Science of Wuhan University 35:491-494.

Georgy J, Noureldin A, Korenberg MJ, Bayoumi MM (2010) Low-cost three-dimensional navigation solution for RISS/GPS integration using mixture particle filter. IEEE Transactions on Vehicular Technology 59:599-615.

Goldberg L (1996) Vision-based vehicle navigation system uses artificial vision to plug GPS's 'blind spots'. Electronic Design 44:35-36.

Golledge R (1995) Path Selection and Route Preference in Human Navigation: A Progress Report. In: Spatial Information Theory-A Theoretical Basis for GIS, vol. 988 (Frank, A. and Kuhn, W, eds), pp 207-222: Springer/Berlin.

Groves PD, Mather CJ (2010) Receiver interface requirements for deep INS/GNSS integration and vector tracking. Journal of Navigation 63:471-489.

Guo Z, Miao LJ, Shen J (2010) Adaptive control of rotary position system in inertial navigation. Beijing Ligong Daxue Xuebao/Transaction of Beijing Institute of Technology 30:552-556+572.

Gupta RA, Masoud AA, Chow MY (2010) A delay-tolerant potential-field-based network implementation of an integrated navigation system. IEEE Transactions on Industrial Electronics 57:769-783.

Haala N, Böhm J (2003) A multi-sensor system for positioning in urban environments. ISPRS Journal of Photogrammetry and Remote Sensing 58:31-42.

Han S, Wang J (2010) Land vehicle navigation with the integration of gps and reduced ins: Performance improvement with velocity aiding. Journal of Navigation 63:153-166.

Hile H, Vedantham R, Cuellar G, Liu A, Gelfand N, Grzeszczuk R, Borriello G (2008) Landmark-based pedestrian navigation from collections of geotagged photos. In: Proceedings of the 7th International Conference on Mobile and Ubiquitous Multimedia (MUM'08), pp 145-152.

Hilton J (2010) Ford motor company and telenav to launch connected in-vehicle GPS navigation system. Automotive Industries AI 190.

Huang C, Liao Z, Zhao L (2010) Synergism of INS and PDR in self-contained pedestrian tracking with a miniature sensor module. IEEE Sensors Journal 10:1349-1359.

Islam A, Iqbal U, Langlois JMP, Noureldin A (2010) Implementation methodology of embedded land vehicle positioning using an integrated GPS and multi sensor system. Integrated Computer-Aided Engineering 17:69-83.

Jeon YW, Daimon T (2010) Study of in-vehicle route guidance systems for improvement of right-side drivers in the Japanese traffic system. International Journal of Automotive Technology 11:417-427.

Jin J, Bian S (2010) Analysis of inertial navigation system positioning error caused by gravity disturbance. Wuhan Daxue Xuebao (Xinxi Kexue Ban)/ Geomatics and Information Science of Wuhan University 35:30-32+41.

Jirawimut R, Prakoonwit S, Cecelja F, Balachandran W (2003) Visual odometer for pedestrian navigation. IEEE Transactions on Instrumentation and Measurement 52:1166-1173.

Jirawimut R, Prakoonwit S, Cecelja F, Balachandran W (2003) Visual odometer for pedestrian navigation. IEEE Transactions on Instrumentation and Measurement 52:1166-1173.

Jo BC, Kim S (2010) Multipath interference cancellation technique for high precision tracking in GNSS receiver. IEICE Transactions on Communications E93-B:1961-1964.

Jong GJ, Horng GJ (2010) Heterogeneous Vehicle Navigation Platform Combined with Mobile RFID Using MIMO System. Wireless Personal Communications 1-12.

Karimi HA (1996) Real-time optimal-route computation: a heuristic approach. Journal of intelligent transportation systems 3 (2):111-127.

Karimi HA, Conahan T, Roongpiboonsopit D (2006) A methodology for predicting performances of map-matching algorithms. In: 6th International Symposium on Web and Wireless Geographical Information Systems (W2GIS 2006), Hong Kong, 4-5 December 2006. Springer, pp 202-213.

Karimi HA, Durcik M, Rasdorf W (2004) Evaluation of uncertainties associated with geocoding techniques. Computer-Aided Civil and Infrastructure Engineering 19 (3):170-185.

Karimi HA, Liu S (2004) Developing an automated procedure for extraction of road data from high-resolution satellite images for Geospatial Information Systems. Journal of Transportation Engineering 130 (4):621-631.

Karimi HA, Sutovsky P, Durcik M (2008) Accuracy and performance assessment of a window-based heuristic algorithm for real-time routing in map-based mobile applications. Map-Based Mobile Services:248-266.

Kasemsuppakom P, Roongpiboonsopit D, Karimi H (2010) Current trends and future direction in GIS. In: Karimi H, Akinci B (eds) CAD and GIS Integration. Auerbach Puplications, Boca Raton, FL, pp 23-49.

Ki Lee J, Jekeli C (2010) Neural network aided adaptive filtering and smoothing for an integrated INS/GPS unexploded ordnance geolocation system. Journal of Navigation 63:251-267.

Kray C, Elting C, Laakso K, Coors V (2003) Presenting route instructions on mobile devices. In: Proceedings of the 8th international conference on Intelligent user interfaces Miami, Florida, USA

Leandro RF, Santos MC, Langley RB (2010) Analyzing GNSS data in precise point positioning software. GPS Solutions 1-13.

Lee WC, Ma MC, Cheng BW (2010) Field comparison of driving performance using a portable navigation system. Journal of Navigation 63:39-50.

Lee YL (2010) The effect of congruency between sound-source location and verbal message semantics of in-vehicle navigation systems. Safety Science 48:708-713.

Lehmann F (2009) Deterministic particle filtering for GPS navigation in the presence of multipath. AEU - International Journal of Electronics and Communications 63:939-949.

Li C (2006) User preferences, information transactions and location-based services: A study of urban pedestrian wayfinding. Computers, Environment and Urban Systems 30:726-740.

Li C, Wang Q, Zhu Z, Guan Z (2007) Study on the GPS/MM navigation method based on the vehicle kinematic model. Yi Qi Yi Biao Xue Bao/Chinese Journal of Scientific Instrument 28:150-153.

Li G, Lu H (2010) Visual recommendation knowledge service based on intelligent topic map. Information Technology Journal 9:1158-1164.

Li J (2010) Improving the accuracy of GPS positioning with an exponential stochastic model. In: 2010 Second International Conference on Networks Security, Wireless Communications and Trusted Computing, vol. 2, pp 162-165 Wuhan, Hubei.

Li JZ, Wu MQ (2010) Three satellites positioning algorithm based on AGPS structure and realization. Dianzi Keji Daxue Xuebao/Journal of the University of Electronic Science and Technology of China 39:372-375+424.

Li T, Yang D, Geng H, Liu W, Zhang T, Lian X (2010) Spider-web road network model for in-vehicle navigation digital maps. Wuhan Ligong Daxue Xuebao (Jiaotong Kexue Yu Gongcheng Ban)/Journal of Wuhan University of Technology (Transportation Science and Engineering) 34:439-442.

Lin CT, Wu HC, Chien TY (2010) Effects of e-map format and sub-windows on driving performance and glance behavior when using an in-vehicle navigation system. International Journal of Industrial Ergonomics 40:330-336.

Liu W, Xu H, Zhang J (2010) Study of high precision and MINS/GPS integrated navigation systems. Wuhan Daxue Xuebao (Xinxi Kexue Ban)/ Geomatics and Information Science of Wuhan University 35:160-163.

Liu ZL, Shi ZK (2010) A dynamic route planning strategy based on road network changes. Jiaotong Yunshu Xitong Gongcheng Yu Xinxi/Journal of Transportation Systems Engineering and Information Technology 10:147-152.

Ning XL, Ma X (2010) Analysis of filtering methods for satellite celestial navigation. Kongzhi Lilun Yu Yinyong/Control Theory and Applications 27:423-430.

Paetzold F, Franke U (2000) Road recognition in urban environment. Image and Vision Computing 18:377-387.

Pang G, Chu MH (2007) Route Selection for Vehicle Navigation and Control. In: The 33rd Annual Conference of the IEEE on Industrial Electronics Society (IECON 2007), pp 2586-2591 Taipei, Thailand.

Pang GaC, M.H. (2007) Route Selection for Vehicle Navigation and Control. In: Industrial Electronics Society (IECON 2007), pp 2586-2591 Taipei, Thailand.

Panga GKH, Takashib, K., Yokotab, T., and Takenagab, H. (2002) Intelligent route selection for in-vehicle navigation systems. Transportation Planning and Technol 25:175-213.

Panzieri S, Pascucci F, Ulivi G (2001) An outdoor navigation system using GPS and inertial platform. In: Proceedings of IEEE/ASME International Conference on Advanced Intelligent Mechatronics, pp 1346-1351 Como, Italy.

Park B, Kee C (2010) The compact network RTK method: An effective solution to reduce GNSS temporal and spatial decorrelation error. Journal of Navigation 63:343-362.

Park SK, Suh YS (2009) Gait state classification by hmms for pedestrian inertial navigation system. Transactions of the Korean Institute of Electrical Engineers 58:1010-1018.

Patino-Studencka L (2010) Approach for detection and identification of multiple faults in satellite navigation. In: IEEE PLANS, Position Location and Navigation Symposium, pp 221-226 Indian Wells/Palm Springs, California.

Patino-Studencka L, Rohmer G, Thielecke J (2010) Researches on the new system of aids to navigation. In: IEEE PLANS, Position Location and Navigation Symposium, pp 221-226 Indian Wells/Palm Springs, California.

Peeta SaY, J.W. (2005) A Hybrid Model for Driver Route Choice Incorporating En-Route Attributes and Real-Time Information Effects. Networks and Spatial Economics 5:21-40.

Petrovski I, Okano K, Ishii M, Torimoto H, Konishi Y, Shibasaki R (2003) Pedestrian ITS in Japan: Pseudolites and GPS. GPS World 14:33-37.

Polat E, Yeasin M, Sharma R (2003) A 2D/3D model-based object tracking framework. Pattern Recognition 36:2127-2141.

Ren C, Pan Y, Liu J, Teng Y (2010) Study on multi-information fusion system to improve the vehicle's position and satellite tracking accuracy. Yingyong Jichu yu Gongcheng Kexue Xuebao/ Journal of Basic Science and Engineering 18:358-367.

Rizos C, Grejner-Brzezinska DA, Toth CK, Dempster AG, Li Y, Politi N, Barnes J, Sun H, Li L (2010) Hybrid positioning a prototype system for navigation in GPS-challenged environments. GPS World 21:42-47.

Roongpiboonsopit D, Karimi H (2010) Comparative evaluation and analysis of online geocoding services. International Journal of Geographical Information Science 24 (7):1081-1100.

Roongpiboonsopit D, Karimi H (2010) Quality assessment of online street and rooftop geocoding services. Cartography and Geographic Information Science 37 (4):301-318.

Roongpiboonsopit D, Karimi HA (2009) A multi-constellations satellite selection algorithm for integrated GNSSs. Journal of Intelligent Transportation Systems 13 (3):127-141.

Sabatini AM (2008) Adaptive filtering algorithms enhance the accuracy of low-cost inertial/ magnetic sensing in pedestrian navigation systems. International Journal of Computational Intelligence and Applications 7:351-361.

Sadeghian P, Kantardzic M, Lozitskiy O, Sheta W (2006) The frequent wayfinding-sequence (FWS) methodology: Finding preferred routes in complex virtual environments. International Journal of Human-Computer Studies 64:356-374.

Saeed G, Brown A, Knight M, Winchester M (2010) Delivery of pedestrian real-time location and routing information to mobile architectural guide. Automation in Construction 19:502-517.

Santa J, Toledo-Moreo R, Zamora-Izquierdo MA, Úbeda B, Gómez-Skarmeta AF (2010) An analysis of communication and navigation issues in collision avoidance support systems. Transportation Research Part C: Emerging Technologies 18:351-366.

Schmid A, Neubauer A, Ehm H, Weigel R, Lemke N, Heinrichs G, Winkel J, Angel Ávila-Rodríguez J, Kaniuth R, Pany T, Eissfeller B, Rohmer G, Overbeck M (2005) Combined Galileo/ GPS architecture for enhanced sensitivity reception. AEU - International Journal of Electronics and Communications 59:297-306.

Schmid F (2008) Knowledge-based wayfinding maps for small display cartography. Journal of Location-based Services 2:57-83.

Schwagten B, Witsenburg M, De Groot NMS, Jordaens L, Szili-Torok T (2010) Effect of Magnetic Navigation System on Procedure Times and Radiation Risk in Children Undergoing Catheter Ablation. American Journal of Cardiology 106:69-72.

Serrano L, Kim D, Langley RB (2010) Multipath adaptive filtering in GNSS/RTK-based machine automation applications. In: Position Location and Navigation Symposium (PLANS), 2010 IEEE/ION, pp 60-69 Indian Wells, CA, USA.

Shen Z, Georgy J, Korenberg MJ, Noureldin A (2010) FOS-based modelling of reduced inertial sensor system errors for 2D vehicular navigation. Electronics Letters 46:298-299.

Shi W, Shen S, Liu Y (2009) Automatic generation of road network map from massive GPS vehicle trajectories. In: 12th International IEEE Conference on Intelligent Transportation Systems (ITSC '09), pp 48-53 St. Louis, MO.

Shin DH, Joo BY (2010) Design of a vision-based autonomous path-tracking control system and experimental validation. Proceedings of the Institution of Mechanical Engineers, Part D: Journal of Automobile Engineering 224:849-864.

Sivaraman S, Trivedi MM (2010) A general active-learning framework for on-road vehicle recognition and tracking. IEEE Transactions on Intelligent Transportation Systems 11:267-276.

St. Amant R, Long T, Dulberg MS (1998) Experimental evaluation of intelligent assistance for navigation. Knowledge-Based Systems 11:61-70.

Stentz A, Hebert M (1995) A complete navigation system for goal acquisition in unknown environments. Autonomous Robots 2:127-145.

Stirling R, Fyfe K, Lachapelle G (2005) Evaluation of a new method of heading estimation for pedestrian dead reckoning using shoe mounted sensors. Journal of Navigation 58:31-45.

Stopher P, FitzGerald C, Zhang J (2008) Search for a global positioning system device to measure person travel. Transportation Research Part C: Emerging Technologies 16:350-369.

Sugimoto C, Nakamura Y, Hashimoto T (2008) Pedestrian-to-vehicle communication system for improving road safety using cellular phones. In: Proceedings of INSS 2008 the 5th International Conference on Networked Sensing Systems, p 226.

Terradellas E, Téllez B (2010) The use of products from ground-based GNSS observations in meteorological nowcasting. Advances in Geosciences 26:77-82.

Tideman M, van der Voort MC, van Arem B (2010) A new scenario based approach for designing driver support systems applied to the design of a lane change support system. Transportation Research Part C: Emerging Technologies 18:247-258.

Toledo-Moreo R, Pinzolas-Prado M, Cano-Izquierdo JM (2010) Maneuver prediction for road vehicles based on a neuro-fuzzy architecture with a low-cost navigation unit. IEEE Transactions on Intelligent Transportation Systems 11:498-504.

Toledo-Moreo R, Zamora-Izquierdo MA (2010) Collision avoidance support in roads with lateral and longitudinal maneuver prediction by fusing GPS/IMU and digital maps. Transportation Research Part C: Emerging Technologies 18:611-625.

Tsuge M, Tokunaga M, Nakano K, Sengoku M (2010) On Estimation of Link Travel Time in Floating Car Systems. International Journal of Intelligent Transportation Systems Research 1-13.

van Eijk O, Roeloffs M (2010) Forensic acquisition and analysis of the Random Access Memory of TomTom GPS navigation systems. Digital Investigation 6:179-188.

Vasconcelos JF, Cardeira B, Silvestre C, Oliveira P, Batista P (2010) Discrete-Time Complementary Filters for Attitude and Position Estimation: Design, Analysis and Experimental Validation. IEEE Transactions on Control Systems Technology.

Wang QT, Hu XL (2010) Improved Kalman filtering algorithm for passive-BD/SINS integrated navigation system based on UKF. Xi Tong Gong Cheng Yu Dian Zi Ji Shu/Systems Engineering and Electronics 32:376-379.

Won SHP, Melek WW, Golnaraghi F (2010) A kalman/particle filter-based position and orientation estimation method using a position sensor/inertial measurement unit hybrid system. IEEE Transactions on Industrial Electronics 57:1787-1798.

Wu Q, Duan F, Shi L, Li J (2010) Design and implementation of a vehicle multimedia navigation terminal software. Wuhan Ligong Daxue Xuebao (Jiaotong Kexue Yu Gongcheng Ban)/Journal of Wuhan University of Technology (Transportation Science and Engineering) 34:573-576.

Xiang L, Liu Y, Su B (2008) New nonlinear filter for GPS/DR integrated navigation system. Dongnan Daxue Xuebao (Ziran Kexue Ban)/Journal of Southeast University (Natural Science Edition) 38:27-31.

Xiang L, Liu Y, Su BK (2010) Improved particle filter algorithm for INS/GPS integrated navigation system. Kongzhi Lilun Yu Yinyong/Control Theory and Applications 27:159-163.

Xu H, Liu H, Tan CW, Bao Y (2010) Development and application of an enhanced kalman filter and global positioning system error-correction approach for improved map-matching. Journal of Intelligent Transportation Systems: Technology, Planning, and Operations 14:27-36.

Xu Z, Li Y, Rizos C, Xu X (2010) Novel hybrid of LS-SVM and kalman filter for GPS/INS integration. Journal of Navigation 63:289-299.

Yanqing W, Deyun C, Chaoxia S, Peidong W (2010) Vision-based road detection by Monte Carlo method. Information Technology Journal 9:481-487.

Zhou J, Knedlik S, Loffeld O (2010) INS/GPS Tightly-coupled integration using adaptive unscented particle filter. Journal of Navigation 63:491-511.

Zhou Z, Shen Y, Li B (2010) A windowing-recursive approach for GPS real-time kinematic positioning. GPS Solutions 1-9.

Zhu W, Boriboonsomsin K, Barth M (2010) Defining a freeway mobility index for roadway navigation. Journal of Intelligent Transportation Systems: Technology, Planning, and Operations 14:37-50.

Chapter 3
Indoor Navigation

3.1 Introduction

From the operational standpoint, indoor navigation systems/services perform similar functions as those in outdoor navigation systems/services. However, they are dissimilar with respect to some of the technologies they use, the physical space within which navigation is performed, and the requirements for routing and directions. The overall similarities and differences between indoor and outdoor navigation systems/services were highlighted in Chapter 1 (see Table 1.1). In this chapter, the characteristics and issues of indoor navigation systems/services are discussed in more detail.

Nowadays, there are technologies that are applicable in both outdoor and indoor navigation systems/services, and there are technologies specifically designed for indoor navigation systems/services. Of the positioning sensors, WiFi and RFIP are becoming widespread in buildings, where they can provide position data for different modes of travel in both outdoors and indoors. However, given the deployment nature of these sensors and the level of accuracy for indoor navigation, a different infrastructure may be considered for each indoor environment.

The main reason why different technologies and different routing and directions are required for outdoor navigation and indoor navigation stems primarily from the differences in the physical spaces which navigation takes place. While in outdoors the physical space for navigation, which consists of road and sidewalk networks, is structured based on standard characteristics, there are no such standard characteristics in indoors. In general, common structures are used in the design of roads, insofar as navigation is concerned, in all geographic areas; for example, roads in different cities and counties have similar structures. In contrast with roads and sidewalks, buildings have different structures (each building may have a different exterior and interior), different heights, different levels (each building may have a different number of floors), and different layouts (each floor of a building may have a different layout).

An important issue that needs to be addressed is the demand for indoor navigation systems/services. While the high demand for outdoor navigation systems/services is easy to explain and justify, the need for indoor navigation systems/services, despite the recent trend in developing indoor navigation systems/services, needs

H. A. Karimi, *Universal Navigation on Smartphones,*
DOI 10.1007/978-1-4419-7741-0_3, © Springer Science+Business Media, LLC 2011

further justifications. It is conceivable that the mobility within buildings is not faced with most of the mobility challenges that are common in outdoors. For example, the physical space in outdoors within which navigation by cars is required may range from a few kilometers to several kilometers to hundreds of kilometers, whereas the physical space in indoors is confined within the geometric extent of the structure of the building and even with the largest structure, the distance between the farthest locations in a building is often a fraction of a kilometer. Other navigation challenges in outdoors include changes in weather and traffic and occurrence of unplanned events such as accident. Such challenges are not of concerns in indoors, though there may be unplanned events such as elevator malfunction which limits mobility (across floors) for a period of time.

With these differences between outdoor and indoor navigation systems/services, it is worth noting that unlike outdoor navigation systems/services where they can be used for a wide range of activities and in different situations, indoor navigation systems/services are meaningful only in special situations. The situations in which indoor navigation assistance is practical include visiting unfamiliar buildings, especially those with complex structures, for first time and mobility during an emergency.

3.2 Technologies

Similar to outdoor navigation systems/services, the main technologies in indoor navigation systems/services include geo-positioning, wireless communication, and database. The main functions performed by indoor navigation systems/services are tracking and routing.

3.2.1 Geo-positioning

Today there are various geo-positioning sensors that can be employed in indoor navigation systems/services. Similar to geo-positioning sensors for outdoor navigation, geo-positioning sensors for indoor navigation can be categorized into absolute positioning and relative positioning.

Table 3.1 shows possible geo-positioning sensors for indoor navigation. These sensors include RFID, WiFi, ultrasound, Bluetooth, image processing, and dead reckoning. While GPS is the most dominant geo-positioning sensor for outdoor navigation, of the available geo-positioning sensors for indoor navigation, RFID

Table 3.1 Geo-positioning sensors for indoor navigation

Sensor	Coverage	Accuracy	Latency	Deployment	Cost to User
WiFi, RFID, ultrasound, Bluetooth, image processing, dead reckoning	Floor	Centimeters to meters	Fast	Infrastructure based	Device (HW/SW), service

Table 3.2 Parameters for indoor navigation

Mode of Travel	Speed*	Route Length**	Structure	Accuracy	Preferences/ Needs	Ambiguous Cases
Walking (Pedestrian)	Slow	Short	Hallway segments, multiple floors	Better than 1 m	Preferences: stairs, elevators	Decision points: multi-ways in hallways
Riding Wheelchair	Slow	Short	Hallway segments, multiple floors	Better than 1 m	Needs: ramps, elevators, accessible restrooms	Decision points: multi-ways in hallways

*Speed: slow < 1 m/sec; **Route Length: fraction of a kilometer

is the most dominant geo-positioning sensor. As shown in Table 3.1, these sensors share similar characteristics when they are used for navigation in indoors. As for coverage, indoor navigation requires that geo-positioning sensors be available on each floor of a building, and depending on the layout of each floor a different infrastructure (i.e., number of units and locations) may be needed. With current geo-positioning sensors, it is possible to obtain accuracies in the range of centimeters to meters, depending on the infrastructure (higher accuracy requires denser infrastructures). There is really no latency issue with these geo-positioning sensors. They are all infrastructure-based and their cost to the user typically includes devices and services.

Table 3.2 shows the parameters that are of importance for navigation in indoors and must be considered in the design and development of indoor navigation systems/ services. These parameters are analyzed based on two main modes of travel, walking and riding wheelchairs. One parameter is speed of the user which in general, whether walking or riding wheelchairs, is slow and almost the same (close to 1 m/sec). Another parameter is length of a route the user must walk or ride in a wheelchair to go from one location to another location of the building. Such trips are often short and even in large buildings may only be a fraction of a kilometer. With respect to structure in both modes of travel, hallway segments and multiple floors are the main considerations. For navigation purposes, both modes of travel require better than 1 m accuracy. However, with respect to preferences and needs, walkers may prefer elevators, whereas wheelchair riders's needs may include ramps, elevators, and accessible restrooms. The main decision points are multi-ways in hallways for both modes of travel.

3.2.2 Wireless Communication

Wireless communication plays a major role in indoor navigation as it is the underlying technology for geo-positioning as well as other activities. While in most cases, wireless communication in outdoor navigation systems/services is for provid-

Table 3.3 Wireless communication systems for indoor navigation

System	Bandwidth	Cost to User
RFID, WiFi, ultrasound, Bluetooth	110 Mb/s	Device, service

ing real-time information, wireless communication in indoor navigation systems/ services is primarily for geo-positioning (i.e., providing position data). Table 3.3 shows the various wireless communication systems (i.e., RFID, WiFi, Ultrasound, and Bluetooth) that are typically used in indoor navigation systems/services. The bandwidth of such systems is in the range of 110 Mb/s and their cost to the user includes devices and services.

3.2.3 Database

The data needed for indoor navigation are different than those needed for outdoor navigation. Table 3.4 shows the types of data needed for indoor navigation. Indoor navigation systems/services require two types of data: hallway segments and hallway networks. Hallway segments in indoors are comparable to road/sidewalk segments in outdoors. However, unlike road/sidewalk segments, hallway segments often have more straightforward geometry (it is reasonable to assume straight lines for most hallway segments). Hallway segments are represented in the vector data model in the database through a series of coordinates that make up each segment shape. A hallway segment may include such attributes as floor number, floor name, and a range of office/room numbers. Hallway segments in indoor navigation systems/services are mainly used for data retrieval (e.g., searching POI) and mapping (e.g., for visual guidance to users).

Hallway networks are another type of data needed for indoor navigation. A hallway network is represented in the vector model through a network (graph) of nodes (decision points) and links (hallway segments). Hallway network data mainly represent topologies of buildings and hallway segment data mainly represent geometries of buildings. There could be three decision points in buildings: multi-way, orientation, and terminal. A multi-way decision point is similar to an intersection in roads/

Table 3.4 Data and database characteristics for indoor navigation

Database	Spatial Data (geometry)	Type	Non-Spatial Data	Functions
Hallway Segment	Series of coordinates on segments making up segment shape	Geometry, vector	Floor number (name), range of office (room) numbers	Retrieval, mapping
Hallway Network	Coordinates of decision points	Topology (decision points), vector	Floor number (name), range of office (room) numbers	Routing, directions

sidewalks where the user needs to make a decision as to which new hallway segment to turn while navigating. An orientation decision point is a point at which user's orientation changes. A terminal point is a point at which there is no other segments to follow. Figure 3.1 shows these decision points in a hallway network. Each hallway network may contain attribute data such as floor number, floor name, and a range of office/room numbers. Hallway networks are primarily for computing routes and directions in indoor navigation systems/services.

In Figure 3.1 a building's layout with the hallway network (dashed lines) and some POIs (e.g., room, drinking fountain, elevator, and restroom) are shown.

Figure 3.2(a) shows a building's layout which contains POIs such as office, lab, and storage space, hallway, and corridor. Figure 3.2(b) shows an example of a route in a building.

Figure 3.3 shows an Entity-Relationship (ER) diagram for navigation databases that provide navigation assistance in indoors with walking as mode of travel.

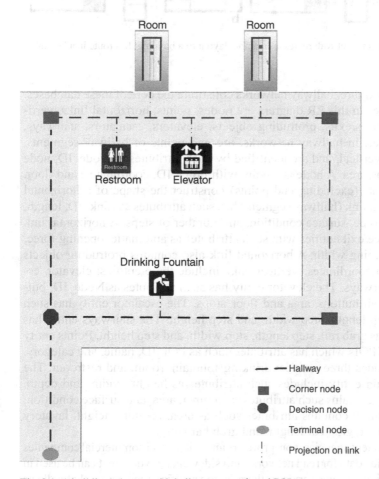

Fig. 3.1 A building's layout with hallway network and POIs

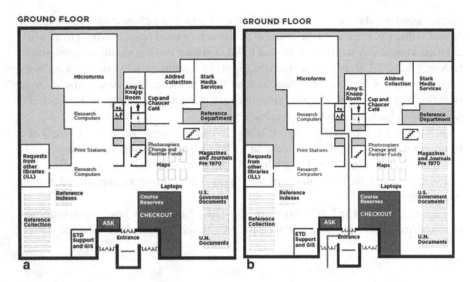

Fig. 3.2 A building's layout with routes. **a** Example layout of a building. **b** A route in a building

As shown in the figure, hallway networks constitute the core of these databases. The main entities in this ER diagram are nodes, points, horizontal links, vertical links, entrances/exits, protruding objects, elevators, escalators, stairways, and POIs. Nodes, in hallway networks, are end points of hallway segments, horizontal and vertical, and are identified by such attributes as Node_ID, node type, and turning area. A node is a point with Point_ID, coordinates, and floor. A series of points (excluding end points) construct the shape of a horizontal link. A horizontal link (hallway segment) has such attributes as Link_ID, length, width, hallway type, surface condition, and number of steps. A horizontal link contains entrance/exit entities with such attributes as automatic, opening force, weight, and opening width. A horizontal link also contains protruding objects with height and coordinates. Vertical links include three entities: elevator, escalator, and stairways. The elevator entity has such attributes as Node_ID, buttons heights, brail buttons, area and floor stops. The escalator entity has such attributes as step length, step width, and step height. The stairways entity has such attributes as grab rail, step length, step width, and step height. Points entity also consists of POIs which has attributes such as POI_ID, name, and category. POI entity includes three entities: drinking fountain, room, and restroom. The drinking fountain entity includes such attributes as height, width, and depth. The room entity contains such attributes as number, area, and surface condition. The restroom entity contains attributes such as area, lavatory height, lavatory depth, toilet height, grab bar length, and grab bar height.

While government agencies, non-profit organizations, and commercial companies collect and provide data for road networks and sidewalk networks that can be used in outdoor navigation systems/services, there is no specific organization that collects

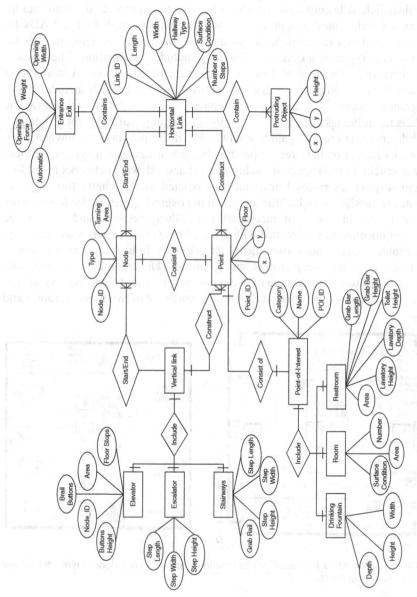

Fig. 3.3 An ER diagram for hallway networks in navigation databases

data for hallway networks suitable for indoor navigation. The data for hallway networks of buildings may be collected and provided by building owners or custodians. However, there are no standards for such data collection and the collected data may not be made available for public use such as in indoor navigation systems/services.

Besides field data collection, which is one common approach to obtain data for hallway networks, another approach is gathering data through CADs. CADs for a building, if they exist, include layout of each floor in the building in high accuracy detail. Typically a CAD for a building includes information either as two-dimensional (e.g., locations of doors) or three-dimensional objects. A more recent approach for obtaining data for hallway networks is BIM which includes building geometry, topology, quantities, and properties. Figure 3.4 shows an example of CADs for buildings and Figure 3.5[1,2] shows an example of BIMs for buildings.

While attention must be paid to quality of hallway networks for navigation in indoors, its impact on the overall operation by indoor navigation systems/services is not as crucial as the impact of quality of road and sidewalk networks in outdoor navigation systems/services. Uncertainties associated with geometry, topology, and attribute are possible but with little impact on the overall operation. Table 3.5 shows different types and sources of uncertainties in hallway networks and POIs. Three sources of uncertainties associated with hallway networks are geometry, topology, and attribute. Uncertainties associated with geometry in hallway networks are of three types: accuracy, completeness, and resolution. Uncertainties associated with topology in hallway networks are of two types: accuracy and completeness. Uncertainties associated with attribute in hallway networks are of two types: accuracy and

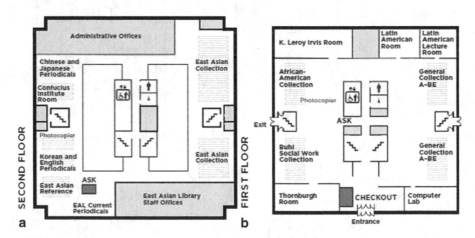

Fig. 3.4 Example CAD. **a** Example CAD for a building and its second floor's layout. **b** Example CAD with floor plan details

---

[1] http://www.knutsonconstruction.com/news/newsletter_articles/building_information_modeling_bim/.

[2] http://www.wsarchitects.com/expertise-bim_ipd.html.

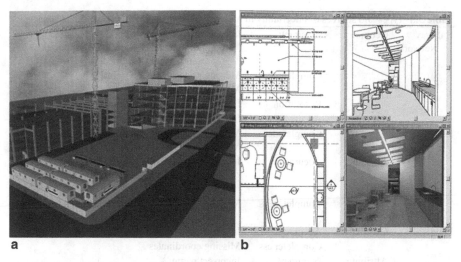

a   b

Fig. 3.5 Example BIM. **a** Example BIM for a building. **b** Example BIM with detailed information

completeness. Two sources of uncertainties associated with POIs are geometry and attribute. Uncertainties associated with geometry in POIs are of two types: accuracy and completeness. Uncertainties associated with attribute in POIs are of two types: accuracy and completeness.

In hallway networks, examples of uncertainties associated with geometry under accuracy type include inaccurate coordinates of nodes (e.g., the location of a decision point stored in the database is far from its true location) and inaccurate coordinates of points representing links (e.g., the points forming a hallway segment do not match the correct location of the hallway segment). Examples of uncertainties associated with geometry under completeness type include missing nodes (e.g., a decision point is not stored in the database) and missing links (e.g., a hallway segment is not stored in the database). An example of uncertainties associated with geometry under resolution type is insufficient points representing segments (e.g., the small number of points on a hallway segment do not represent the true shape of the hallway segment).

In hallway networks, an example of uncertainties associated with topology under accuracy type is inaccurate coordinates of nodes (e.g., a decision point does not have the correct connection to hallway segments). Examples of uncertainties associated with topology under completeness type is missing nodes (e.g., a decision point is not stored in the database) and missing links (e.g., a hallway segment is not stored in the database).

In hallway networks, an example of uncertainties associated with attribute under accuracy type is incorrect segment accessibility information (e.g., inaccurate hallway segment width, inaccurate hallway segment length, or inaccurate hallway segment surface). An example of uncertainties associated with attribute under completeness type is missing accessibility information (e.g., segment width, segment length, or segment surface is not stored in the database).

Table 3.5 Types and sources of uncertainty in hallway networks and POIs

Hallway Network	Geometry	Accuracy	Inaccurate coordinates of nodes (decision points); Inaccurate coordinates of points representing links (segments)
		Completeness	Missing nodes (decision points); Missing links (segments)
		Resolution	Insufficient points representing segments
	Topology	Accuracy	Incorrect connectivity at decision points
		Completeness	Missing nodes (decision points); Missing links (segments)
	Attribute	Accuracy	Incorrect accessibility information (e.g., segment width, segment length, segment surface)
		Completeness	Missing accessibility information (e.g., segment width, segment length, segment surface)
POIs	Geometry	Accuracy	Incorrect coordinates
		Completeness	Missing coordinates
	Attribute	Accuracy	Incorrect name, type, number, accessibility information
		Completeness	Missing rooms/offices, restrooms, drinking fountain; Missing number, accessibility information

In POIs, an example of uncertainties associated with geometry under accuracy type is incorrect coordinates (e.g., the location of a POI is far from its true location); an example of uncertainties associated with geometry under completeness type is missing coordinates (e.g., the location of a POI is not stored in the database).

In POIs, examples of uncertainties associated with attribute under accuracy type include incorrect POI name, POI type, POI number, and accessibility information (e.g., inaccurate location of an elevator); examples of uncertainties associated with attribute under completeness type include missing rooms/offices, restrooms, drinking fountain, and number and accessibility information.

3.3 Functions

In general, indoor navigation systems/services have similar functions as those in outdoor navigation systems/services. Table 3.6 shows the main functions performed by indoor navigation systems/services.

Retrieval. The retrieval function is responsible for retrieval of data from the database which contains spatial and attribute data. The input to the retrieval function is usually a POI name and the output, depending on the input, could be location of a POI (or a decision point).

Table 3.6 Main functions performed by indoor navigation systems/services

Computation	Input	Process	Output	Database
Retrieval	POI name	Retrieve spatial, non-spatial data	Locations of POI (decision points)	Spatial, non-spatial, attribute
Map Creation	Current location	Determine hallway segment and floor	Map	Hallway segments, non-spatial, attribute
Mapping	Current location	Zoom in, zoom out, pan	Map	Hallway segments, non-spatial, attribute
Geocoding	POI address (e.g., room number)	Interpolation	Location on map	Hallway segments, non-spatial
Routing/Rerouting	Origin-Destination addresses (current location for rerouting)	Optimal routing	Route on map	Hallway networks
Tracking	Geo-positioning data	Estimating current location on hallway segment	Current location on map	Hallway networks, hallway segments, non-spatial
Direction	Route	Compute distance and search for landmarks	A set of instructions to navigate from origin (or current location) to destination	Hallway networks, hallway segments, non-spatial

Map creation. The map creation function is responsible for creating a map using a POI location or current location (obtained through geo-positioning sensors). This function needs all types of data in the database including hallway networks, hallway segments and attribute data.

Mapping. Once a map is created, the user is able to perform mapping functions. The mapping function allows user to zoom in, zoom out, and pan the created map. Similar to the map creation function, the input to the mapping function could be a POI location or current location and the output is a new map. All types of data in the database including hallway networks, hallway segments, and attribute data are usually needed in this function.

Geocoding. The geocoding function is responsible for computing the coordinates of an address or a point. The input to the geocoding function often is a POI address and the output is the location of the POI address on the map. The geocoding process in most cases involves an interpolation scheme that uses spatial information on the end nodes of hallway segments; the geometry of the segment (series of coordinates forming the shape of the segment); and the room numbers on both sides of the segment to estimate the location of a given address. The data needed in the geocoding function include hallway segments and attribute data.

Routing/Re-routing. The routing function is responsible for computing optimal, based on a pre-determined criterion, routes between pairs of origin-destination addresses. An address entered by the user or user's current location (obtained through geo-positioning sensors) can be used as the origin. The output is the computed route highlighted on the map. The main data needed in the routing function is hallway networks which provide the topology of the network. In indoor navigation systems/ services, it is reasonable to assume that the main criterion for routing is shortest distance. An extension to the routing function is rerouting, which depending on the situation (e.g., deviation from the computed route), will re-compute a new route between the current location of the user, as the origin, and the destination.

Tracking. The tracking function is responsible for the continuous monitoring of user's location in real time. The input to the tracking function is the position data acquired continuously at a fixed interval (time or distance) where it is used for map matching. The output is the real-time location of the user displayed on the map. The data used in the tracking function include hallway networks, hallway segments, and attribute data. The process of tracking involves continuously estimating the precise location of the user on the hallway segment.

Direction. The direction function uses the computed route to provide instructions to travel from the origin on each segment of the route to reach the destination. The input to the direction function is a route and the output is a set of instructions to navigate from the origin (or current location) to the destination presented on the map and/or through voice. The set of instructions basically utilizes information on decision points, distances on hallway segments, and landmarks. The data needed for the direction information include hallway networks, hallway segments, and attribute data.

3.4 User

Similar to outdoor navigation systems/services, in this section we discuss usability in indoor navigation systems/services. For an indoor navigation system/service to be able to provide appropriate guidance, it is required that it supports various features such as map representation, navigation situations, purpose of trip, and user preferences. Table 3.7 shows these usability features for different modes of travel in indoor navigation systems/services.

In Table 3.6, users are categorized based on mode of travel (i.e., walking or riding wheelchair). For each group, map content, purpose of trip, and user preferences with respect to POI, route, and map presentation are analyzed.

Map content: For pedestrians, hallway segments and networks and relevant objects (e.g., landmarks) on or around hallway segments need to be presented. For wheelchair riders, hallway segments and networks and relevant objects (e.g., landmarks and inaccessible routes) must be presented. Due to the impediments in passing inaccessible routes, such as steps and obstacles by wheelchair riders, the map must either not present them at all or present them in such a way that are not confusing to wheelchair riders.

Purpose of trip: Users of indoor navigation systems/services are either commuters (those who travel frequently between POIs) or non-commuters (those who travel infrequently between POIs). In general, in indoor navigation systems/services and with any mode of travel, the routing criterion is shortest distance. However, additional criteria for wheelchair riders are least turns, no obstacles and no steps. POIs for both walkers and wheelchair riders are offices/rooms and restrooms. As for map preferences of walkers, voice/text directions, color, brightness, screen size, font size, map scale, and presentation of upcoming POIs are important. As for map preferences of wheelchair riders, voice/text directions, color, brightness, screen size, font size, map scale, presentation of upcoming POIs, and presentation of inaccessible routes with different colors are important.

Table 3.7 Usability features in indoor navigation systems/services

Mode of Travel	Map Content	Purpose of Trip	User Preferences		
			Route	POI	Map Presentation
Pedestrian	Hallways, landmarks	Commute	Shortest path	Office, restroom	Voice/text directions; Color; Brightness; Screen size; Font size; Map scale; selected POIs
		Non-commute	Shortest path	Office, restroom	
Wheelchair Rider	Hallway, landmarks	Commute	Shortest path; Least turns; Avoid obstacles; Avoid steps	Office, restrooms	Voice/text directions; Color; Brightness; Screen size; Font size; Map scale; Selected POIs; Inaccessible routes

3.5 Summary

In this chapter, characteristics, technologies, and techniques of indoor navigation systems/services are discussed. The main technologies in navigation systems/services are geo-positioning, wireless communication, and database. Wireless communication plays a pivotal and different role in indoor navigation systems/services than it does in outdoor navigation systems/services. Of the possible geo-positioning sensors RFID and WiFi are the main geo-positioning sensors for indoor navigation. Characteristics of RFID, WiFi, and other possible geo-positioning sensors for indoor navigation are described. Hallway networks, composed of hallway segments, constitute the core of data in outdoor navigation systems/services. The main functions performed by today's indoor navigation systems/services are POI retrieval, map creation, geocoding, routing/re-routing, and tracking.

Further Readings

Anagnostopoulos C, Tsetsos, V., Kikiras, P., Hadjiefthymiades, S. (2005) OntoNav: A Semantic Indoor Navigation System. In: Proceedings of the 1st Workshop on Semantics in Mobile Environments (SME), vol., Cyprus.

Barnes J, Rizos, C., Wang, J., Small, D., Voigt, G., and Gambale, N. (2003) High Precision Indoor and Outdoor Positioning using LocataNet. Global Positioning Systems 2:9.

Birgri E (2007) Pedestrian Navigation -Creating a Tailored Geodatabase for Routing. Positioning. In: Navigation and Communication (WPNC '07).

Broumandan A, Nielsen J, Lachapelle G (2010) Enhanced detection performance of indoor GNSS signals based on synthetic aperture. IEEE Transactions on Vehicular Technology 59:2711-2724.

Bshara M, Orguner U, Gustafsson F, Van Biesen L (2010) Fingerprinting localization in wireless networks based on received-signal-strength measurements: A case study on wimax networks. IEEE Transactions on Vehicular Technology 59:283-294.

Butz A. B, J., Krüger, A., Lohse, M. (2001) A hybrid indoor navigation system. In: Proceedings of the 6th international conference on Intelligent user interfaces, pp 25-32 Santa Fe, New Mexico, USA.

Cheok AD, Yue L (2010) A Novel Light-Sensor-Based Information Transmission System for Indoor Positioning and Navigation. IEEE Transactions on Instrumentation and Measurement.

Chiou CK, Tseng JCR, Hwang GJ, Heller S (2010) An adaptive navigation support system for conducting context-aware ubiquitous learning in museums. Computers and Education 55:834-845.

Duckham MK, L. (2003) "Simplest" Paths: Automated Route Selection for Navigation. In: COSIT 2003 Spatial information theory (Kuhn, W., Worboys, M.F., Timpf, S., ed), pp 169-185 Berlin: Springer.

Dudas P, Ghafourian, M., Karimi, H.A. (2009) ONALIN: Ontology and Algorithm for Indoor Routing. In: The 1st International Workshop on Indoor Navigation Awareness Taipei, Taiwan.

Fischer C, Muthukrishnan K, Hazas M, Gellersen H (2008) Ultrasound-aided pedestrian dead reckoning for indoor navigation. In: Proceedings of the Annual International Conference on Mobile Computing and Networking (MOBICOM), pp 31-36.

Gilliéron P, Büchel, D., Spassov, I., Merminod, B. (2004) Indoor Navigation Performance Analysis. In: Proceedings of the European Navigation Conference GNSS, pp 17-19 Rotterdam, Netherlands.

Han D, Lee M, Chang L, Yang H (2010) Open radio map based indoor navigation system. In: 8th IEEE International Conference on Pervasive Computing and Communications Workshops (PERCOM Workshops) pp 844-846 Mannheim.

Holscher C, Meilinger T, Vrachliotis G, Brosamle M, Knauff M (2006) Up the Down Staircase: Wayfinding Strategies and Multi-Level Buildings. Journal of Environmental Psychology 26:284-299.

Huang H, Gartner G, Schmidt M, Li Y (2009) Smart environment for ubiquitous indoor navigation. In: Proceedings - 2009 International Conference on New Trends in Information and Service Science, NISS 2009, art no 5260410, pp 176-180 pp 176-180.

Inoune Y, Ikeda, T., Yamamoto, K., Yamashita, T., Sashima, A., Kurumatani, K. (2008) Usability Study of Indoor Mobile Navigation System in Commercial Facilities. In: Proceeding of the 2nd International Workshop on Ubiquitous Systems Evaluation (USE 2008) in UbiComp 2008, pp 45-50 Seoul, South Korea.

Jan SS, Hsu LT, Tsai WM (2010) Development of an indoor location based service test bed and geographic information system with a wireless sensor network. Sensors 10:2957-2974.

Jin Y, Motani M, Soh WS, Zhang J (2010) SparseTrack: Enhancing indoor pedestrian tracking with sparse infrastructure support.

Kargl F, Gebler, S., Flerlage, F. (2007) The iNAV indoor Navigation System. In: Symposium on Ubiquitous Computing Systems (UCS 2007), vol. 4836/2008 Tokyo, Japan.

Karimi HA, Ghafourian M (2010) Indoor Routing for Individuals with Special Needs and Preferences. Transactions in GIS 14:299-329.

Moeser S (1998) Cognitive mapping in a complex building. Environment and Behavior 20:895-913.

Ng W (1994) An integrated electron-beam probing environment with a simulation interface and CAD navigation. Microelectronic Engineering 24:287-294.

Papataxiarhis V, Riga, V., Nomikos, V., Sekkas, O., Kolomvatsos, K., Tsetsos, V., Papageorgas, P., Vourakis, S., Xouris, V., Hadjiefthymiades, S., Kouroupetroglou, G. (2008) MNISIKLIS: Indoor Location Based Services for All. (Gartner, G., Rehrl, K., ed), pp 263-282 Berlin Heidelberg: Springer.

Rajamäki J, Viinikainen P, Tuomisto J, Sederholm T, Säämänen M (2007) LaureaPOP personalized indoor navigation service for specific user groups in a WLAN environment. WSEAS Transactions on Communications 6:524-531.

Raubal M, Egenhofer M (1988) Comparing the complexity of wayfinding tasks in built environments. Environment and Planning B 25:895-913.

Swobodzinski M, M. Raubal. (2009) An indoor routing algorithm for the blind: development and comparison to a routing algorithm for the sighted. International Journal of Geographical Information Science.

Tomono MaY, S. (2002) Indoor Navigation based on an Inaccurate Map using Object Recognition. In: Proceeding of International Conference on Intelligent Robots and Systems (2002 IEEE/RSJ), pp 619-624 EPFL, Switzerland.

Tsetsos V, Anagnostopoulos, C., Kikiras, P., and Hadjiefthymiades, S. (2006) Semantically Enriched Navigation for Indoor Environments. International Journal of Web and Grid Services (IJWGS) 4.

Yaun LaZ, H. (2008) 3D Indoor Navigation: a Framework of Combining BIM with 3D GIS. In: 44th ISOCARP Congress.

Yu XS, Zhao W, Liu P, Tang XL (2010) Estimating the pedestrian 3D motion indoor via hybrid tracking model. Zidonghua Xuebao/Acta Automatica Sinica 36:773-784.

Chapter 4
Universal Navigation

4.1 Introduction

In this book, universal navigation is defined as an environment that provides any users, regardless of location, purpose of trip, special needs and preferences, with personalized navigation assistance. As shown in Figures 1.4 and 1.8, universal navigation is considered as the fourth generation of outdoor navigation technology and the second generation of indoor navigation technology, respectively.

The primary purpose of universal navigation, called Universal NAVIgation Technology (UNAVIT), is to overcome the shortcomings of current navigation systems/services. To better understand the benefits and features of UNAVIT, the shortcomings of current navigation systems/services and how they can be overcome by UNAVIT will be discussed. Table 4.1 presents those questions, reformulated for UNAVIT, and provides responses, as UNAVIT's capabilities.

4.2 Ontology

To better understand the features and capabilities of UNAVIT, in this section we present and discuss an ontology for UNAVIT. Figure 4.1 shows an ontology which contains "user", "request", "recommend", "profile", "preferences", and "adapt" as the main concepts.

The "user" concept in this ontology has relationship with all other concepts. It has the "send" relationship with "request" and "recommend", has the "has" relationship with "profile" and "preferences", and has the "to" relationship with "adapt".

The "profile" concept is for realizing the needs of individuals for navigation assistance and has four sub-concepts, namely "visually impaired", "mobility impaired", "cognitively impaired", and "elderly", all related to the "profile" concept via the "is" relationship.

The "preferences" concept is for realizing the preferences of individuals for routing and has the "contain" relationship with four sub-concepts, namely "safe route",

H. A. Karimi, *Universal Navigation on Smartphones*, 75
DOI 10.1007/978-1-4419-7741-0_4, © Springer Science+Business Media, LLC 2011

Table 4.1 UNAVIT's capabilities

Question	Comments
Can UNAVIT meet the specific needs of all individuals including those with mobility, visual, and cognitive impairments?	UNAVIT will take into account the special needs of individuals with mobility, visual, and cognitive impairments.
Will UNAVIT be adaptable to users with different cognitive abilities, levels of computing knowledge and exposure to technology?	UNAVIT will include features that adapt to individuals with different cognitive abilities and alleviate the need for special computing and technology knowledge to request and receive navigation assistance.
Can UNAVIT provide navigation assistance seamlessly between indoors and outdoors?	UNAVIT will be capable of allowing seamless transition from outdoor to indoor and from indoor to outdoor.
Can UNAVIT support users with navigation assistance suitable for various situations, such as difficult versus easy routes, day versus night, routes on snowy days versus sunny days?	UNAVIT will be able to recognize the differences in navigation and provide appropriate assistance in each situation.
Can UNAVIT provide individuals with proper navigation assistance seamlessly based on the context?	UNAVIT will be context-aware allowing users to receive assistance relevant to their needs, at different locations and times.
Can UNAVIT seamlessly adjust to different modes of travel, such as driving cars, walking, riding bicycles, riding wheelchairs, or Segways?	UNAVIT will be able to recognize the change in mode of travel and adjust to the new environment accordingly.
Can UNAVIT effectively be used in different countries with dissimilar policies and cultures?	UNAVIT will be able to recognize navigation differences with respect to policies and cultures in different countries and provide appropriate assistance in each.

"shortest distance", "shortest time", and "avoid toll"; these sub-concepts are the common criteria for routing in navigation systems/services.

The "request" concept, which could be both for indoor and outdoor navigation, is for the purpose of users requesting navigation assistance and consists of the following sub-concepts: "member's experience in a region", "map mark", "geotag", "locate nearby friends", "locate a friend", "route/direction", and "nearby POI". These sub-concepts are related to the "request" concept via the "is-a' relationship.

The "recommend" concept, which could be both for indoor and outdoor navigation, is for the purpose of recommending navigation activities to users and overlaps with some of the sub-concepts in the "request" concept and includes "locate nearby friends", "locate a friend", "route/direction", and "nearby POI". Similar to the sub-concepts in the "request" concept, these sub-concepts have the "is-a" relationship with the "recommend" concept.

The "adapt" concept is for the purpose of adapting to different navigation situations and is related to five main sub-concepts via the "apply-for" relationship. These sub-concepts are: "time/season changes", "user preferences & needs", "different location-based culture", "travel modes" and "tracking". The "tracking" sub-concept

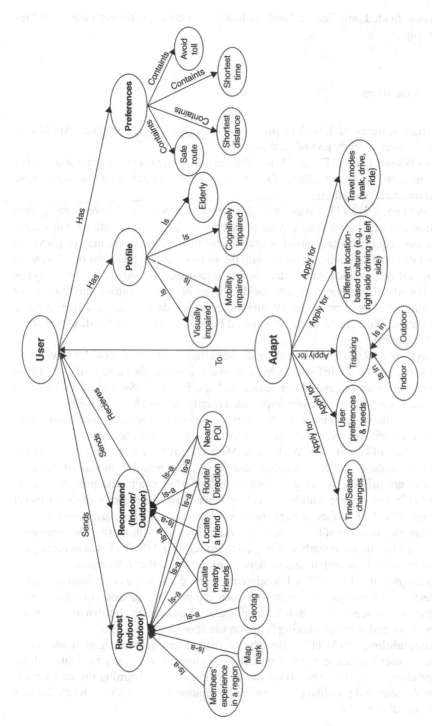

Fig. 4.1 An ontology for UNAVIT

is further divided into "indoor" and "outdoor" sub-concepts through the "is-in" relationship.

4.3 Features

The main features of UNAVIT are AnyWhere, AnyTime, AnyUser, AnyMode-OfTravel, transparency and adaptability.

AnyWhere. UNAVIT's AnyWhere feature allows the user to request and receive navigation assistance regardless of location, in outdoor or in indoor (i.e., navigation based on spatial variation).

AnyTime. UNAVIT's AnyTime feature recognizes the different navigation needs at different times (i.e., navigation based on temporal variation). For example, a user requesting navigation assistance during daylight hours may be provided a route that is different than a route, with the same origin and destination addresses, that would be provided during dark hours; one reason for this could be the types of landmarks that are appropriate for directions on a route during daylight hours which may not be appropriate for directions during dark hours (a different route that contains illuminated landmarks would be more appropriate for directions during dark hours).

AnyUser. UNAVIT's AnyUser (or personalization) feature provides navigation assistance that is suitable for different groups and individuals. Groups of users could be categorized under general population and special needs. Special needs could be sub-categorized under mobility impaired, visually impaired, cognitively impaired, and elderly. Individuals under each group (i.e., general or special needs) may have their preferences for navigation that UNAVIT must take into account.

AnyModeofTravel. UNAVIT's AnyModeOfTravel feature allows transition from one mode of travel to another seamlessly. The possible modes of travel in outdoors are driving, walking, biking, or riding wheelchair. The possible modes of travel in indoors are walking or riding wheelchair. As a user's mode of travel changes, UNAVIT makes the appropriate adjustments without user intervention. For example, as mode of travel changes from driving to walking (the user may park the car at a location and starts walking on the sidewalk), UNAVIT will make adjustments from road network to sidewalk network, among other adjustments.

Transparency. UNAVIT's Transparency feature minimizes user intervention in interacting with navigation systems/services and performing tasks. In other words, navigation is processed in a simple and practical manner so the burden of most navigation tasks and decision making is not on the shoulder of the user.

Adaptability. UNAVIT's Adaptability feature provides navigation assistance based on user's needs as navigation behavior changes. As user's navigational behavior changes, so does the navigation system/service by learning the new behavior of the user and providing appropriate navigation assistance that meets the new navigational behavior.

4.4 Architecture

Figure 4.2 shows an architecture for UNAVIT. The main components of this architecture include smartphones, Kiosks, Meta Navigation (MetaNav), Query Processing Engine (QPE), and Navigation Web Service Providers (NavWSPs).

Due to the advancements in mobile computing and wireless networks, today's smartphones are increasingly turning into PDAs as they can support fairly complex computational tasks, large storage capacity, and memory. Furthermore, smartphones, equipped with GPS and other geo-positioning sensors are ideal gateways to UNAVIT.

Each kiosk in UNAVIT is responsible for maintaining the spatial and attribute data within a given geographic area (e.g., city or county). A large area (e.g., a region or country) will have multiple kiosks where together they contain all the spatial and

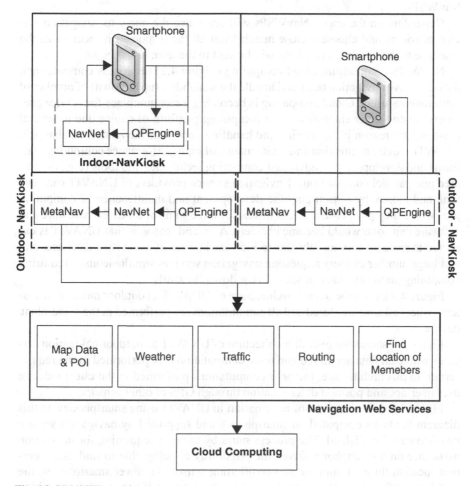

Fig. 4.2 UNAVIT's architecture

attribute data for the area. UNAVIT will feature two types of kiosks: Outdoor-Kiosk and Indoor-Kiosk. Each Outdoor-Kiosk is responsible for maintaining the spatial and attribute data for a specific geographic area. Each Indoor-Kiosk is responsible for maintaining the spatial and attribute data for a building. Depending on user's location, UNAVIT will access the appropriate kiosks.

MetaNav serves as a directory of data available in all kiosks. Once a request, which requires multiple kiosks, is submitted and the corresponding kiosk (the one within which the user is located) is identified, MetaNav will be consulted to find out the other kiosks that are needed to obtain the required and relevant spatial and attribute data.

Once the required kiosks are determined, the query will be submitted to QPE, which is responsible for interpreting and analyzing the query so that the appropriate response can be made. Upon realizing an appropriate response, the query (probably in a form different than how it was originally submitted) will be passed on to NavWSPs.

Depending on the query, NavWSPs will announce the query to navigation service providers and choose a close match from all the responses it receives as the response to the user's query which will be sent to the user.

NavWSPs could utilize cloud computing (Figure 4.2) to process compute- and data-intensive navigation tasks and handle the scalability (up or down) of number of simultaneous users. Cloud computing is becoming a commonplace for service providers, among other enterprises, as a computing platform of choice due to several reasons. One reason is that storing and handling very large navigation databases in UNAVIT require acquisition and maintenance of high-performance computing platforms and development of advanced computing techniques and tools (e.g., distributed and parallel computation). Navigation service providers of UNAVIT can save time and money by outsourcing the development and maintenance of computing jobs where they pay per-usage fees instead of large sums of funds for hardware and software that soon would become obsolete. A second reason is that UNAVIT is expected to scale up very rapidly, due to increase in sizes of navigation databases and to a large number of users requesting navigation services simultaneously, requiring computing platforms that can scale up (or down) instantly.

Figure 4.3 shows the overall architecture of UNAVIT as outdoor navigation systems where all data are stored and all computations are performed in the stand-alone device.

Figure 4.4 shows the overall architecture of UNAVIT as outdoor navigation services where all data are stored and most computations are performed in the remote server. In this architecture, the only computations performed in the client side are user interface and position determination through GPS or other sensors.

Figure 4.5 shows the flow of information in UNAVIT using smartphones. In this diagram the tasks computed on smartphones and supported by navigation service providers are highlighted. The process starts by the user requesting for navigation assistance on a smartphone. Given the possibility of being able to find user's current location through multiple geo-positioning sensors in newer smartphones, the first task performed in the smartphone is to check to see if it is receiving position data through multiple geo-positioning sensors. If only one geo-positioning sensor

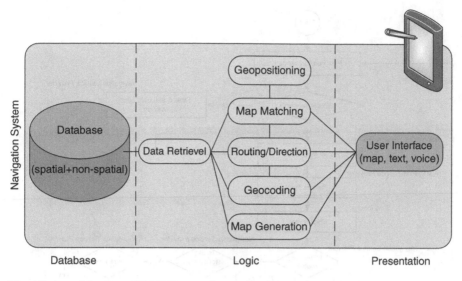

Fig. 4.3 An architecture of UNAVIT as outdoor navigation systems

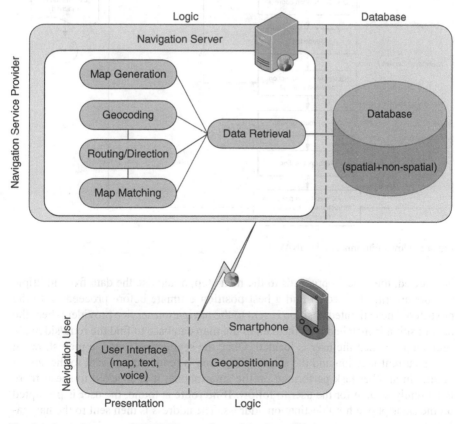

Fig. 4.4 An architecture of UNAVIT as outdoor navigation services

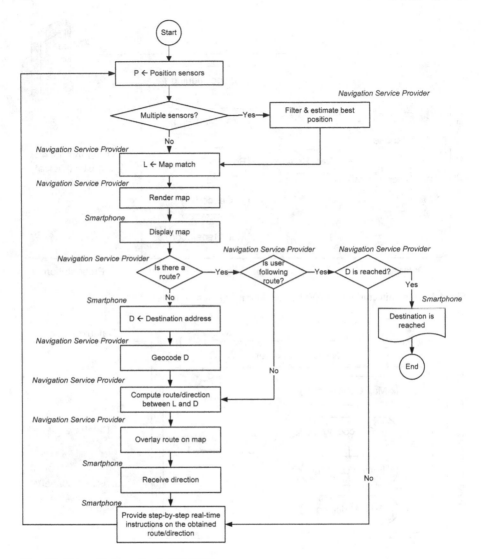

Fig. 4.5 Flow of information in UNAVIT

is detected, the process proceeds to the next step, otherwise the data from multiple sensors are first filtered to find a best position estimate before proceeding to the next step. The estimated position is sent to the navigation service provider where the new position is matched with the data in the map database to find the road/sidewalk segment on which the user is located. Once the segment is found, a map showing user's current location and the surrounding area is rendered and sent to the smartphone. In another task performed on the smartphone, it is checked to see if there is currently a route for the user to follow. If no route is found, the user is prompted on the smartphone for a destination address. The address is then sent to the navigation service provider for geocoding and computing an optimal route (based on a pre-selected criterion) and the direction on it. The computed route and the direction are overlaid on a map which is sent to the smartphone, where the user will receive step-by-step instructions on how to navigate on the computed route. From this point on, the process of receiving position data, map matching, checking for route, geocoding, routing, mapping, and providing direction is repeated. If a route is provided, but the user is not following it, rerouting will be invoked. Rerouting occurs when deviation of travel on a route is detected and involves computing a new route and the direction on it between user's current location and the destination. If a route is provided and the user is following it, then the process will check to see if the destination is reached. If the destination is not reached yet, the process of tracking will continue, otherwise, if the destination is reached, the tracking process for that route is completed.

Since UNAVIT, as new navigation services, are expected to be increasingly available on smartphones and currently Android (from Google) and iPhone (from Apple) are two of the most predominant platforms for smartphones, in Table 4.2, some of the features and capabilities of these two platforms are highlighted. While there are similarities between the two platforms, they are differences in features and capabilities between the two platforms. One major difference between the two platforms, which mainly concerns application developers for smartphones, is that the Android platform is open source while the iPhone platform is not. Another dissimilar feature is that Android heavily relies on its Google resources such as Google Maps data and functions whereas the iPhone uses Google Maps and Yahoo Maps as external resources for application development. As for development cost, Android, as an open source platform, does not require any cost associated with Software Development Kit (SDK) and licensing whereas for development on iPhone there is a fee and licensing agreement. It should be noted that these features and capabilities in Table 4.2 are valid only as of this writing and like any other modern technologies, these platforms are expected to improve and change over time.

WMSs are online services that provide a variety of functions such as mapping, geocoding, and routing and have become very popular in recent years. Examples of WMSs include Google Maps, Bing Maps, and Yahoo Maps. Given the popularity and the demand for WMSs, they are increasingly being used as the underpinning platform for navigation services. Table 4.3 summaries the features and capabilities of Google Maps, Yahoo Maps, MapQuest, Bing Maps, MSN Map, Ask MAP, NAVTEQ, and Wikimapia. Geocoding, displaying points, distance information, route criteria, manual placement of origin and destination locations, and routes and directions between pair of locations are the features and capabilities of these WMSs as summarized in Table 4.3. This table highlights those features and capabilities that are of interest to end users. As for features and capabilities for developers, some WMSs allow access to their Application Programming Interfaces (APIs) while some do not. Having access to APIs facilitates development of new applications on top of the underlying WMSs. However, as of this writing some WMSs, while allow use of their APIs, restrict use of some of their functions.

Table 4.2 Features and capabilities of Android versus iPhone

Feature		Android	iPhone
Developer Environment		Eclipse	Xcode
Programming Languages and Features		Java	Objective-C
		Statically typed	Dynamically typed
		Garbage collection	No garbage collection
Platform		Open-source	Proprietary
Functionality	Mapping	Google: zooming and panning with touch screen	deCarta: zooming and panning with touch screen
	Proximity	Not supported	Supported
	Nearest	Not supported	Supported
	Routing	Google Maps APIs	deCarta
	Others	Overlay	Overlay
		Social networking	Social networking
		Geocoding and reverse geocoding using Google tools	Geocoding and reverse geocoding using Yahoo tools
External Resources	Data	Google Maps	Google Maps
			Yahoo Map
	Functions	Google	deCarta
Geo-positioning	Technology	GPS, WiFi, Cellular triangulation	WiFi, Cellular triangulation, GPS, Skyhook wireless
	Accuracy	N/A	Reasonably accurate for indoor
Computational Capability		Depends on the device	High (up to 4096 FFT points is less than 60 m/sec)
Cost		Free	No deployment: Free Deployment: Standard program: 99$; Enterprise program: 299$
Interoperability		Windows	Mac
		Mac OS X	Windows
		Linux (i386)	

Navigation services are currently being offered by some vendors and their presence in the market is expected to rapidly increase (all current and new smartphones feature navigation services). Table 4.4 shows current navigation services offered by Apple, Google, Nokia, and Blackberry.

Through third-party resources, such as MapQuest 4 Mobile, Apple's iPhone can display current location and real-time traffic and provide routes and directions for drivers and pedestrians via text and voice with criteria such as shortest time and shortest distance and no highways, no tolls and no seasonally closed roads. It can re-route drivers if they deviate from computed routes. It facilitates searching for POIs, supports an energy saving option, and allows for multi-point destination. Through third-party resources such as TomTom, CoPilot Live, and MobileNavi-

Table 4.3 Features and capabilities of Web Mapping Services

	Geocode O,D Addresses	Display Points using X,Y Coordinates	Route/ Direction	Distance Information	Route Criteria	Manual Placement of O and D	Routes/ Directions between O & D using X,Y Coordinates
Google Maps	✓	✓	✓	✓	Driving; Walking; Avoid highway; Avoid toll	✓	✓
Yahoo Maps	✓	✓	✓	✓	Not specified	X	✓
MapQuest	✓	X	✓	✓	Shortest time; Shortest distance; Avoid toll; Avoid highway; Avoid seasonally closed roads	✓	X
Bing Maps	✓	✓	✓	✓	Shortest time; Shortest distance; Route based on traffic	✓	X
MSN Map	✓	X	✓	✓	Shortest time; Shortest distance	X	X
Ask Map	✓	✓	✓	✓	Driving; Walking	X	✓
NAVTEQ	✓	X	✓	✓	Driving; Walking; Shortest time; Shortest distance; Avoid toll; Avoid highway; Avoid ferries	X	X
Wikimapia	X	✓	X	X	Not specified	X	X

gator, Apple's iPhone can work offline, provide routes and directions via text and voice, provide multi-point destination, and display speed limits. Through third-party resources such as AT&T Navigator, Apple's iPhone can provide routes and directions via text and voice and display speed limits. Through third-party resources such as Gokivo Navigator, Apple's iPhone can provide routes and directions via text and voice.

Google's Android uses Google resources to display current location and real-time traffic. It can provide routes and directions for drivers and pedestrians via text and voice, re-route if the driver deviates from the computed route, and provide bus

Table 4.4 Navigation service providers

Company	Product Name/Model	General Characteristics providers	Third-Party Company	URL
Apple	iPhone	Display current location; Real-time traffic; Routes and directions for drivers and pedestrians via text and voice; Shortest time and shortest distance and avoid highways, tolls, and seasonally closed roads; Re-route if deviates from computed route; Search for POIs; Energy saving option; Multi-point destination	MapQuest 4 Mobile	http://mashable.com/2010/03/30/mapquest-iphone/
		Works offline; Routes and directions via text and voice; Multi-point destination; Display speed limits	TomTom; CoPilot Live; Navigon MobileNavigator	http://www.engadget.com/2010/04/09/iphone-navigation-shootout/
		Routes and directions via text and voice; Display speed limits	AT&T Navigator Gokivo Navigator	
		Routes and directions via text and voice		
Google	Android	Display current location; Real-time traffic; Routes and directions for drivers and pedestrians via text and voice; Re-route if driver deviates from computed route; Bus options based on nearby bus stops; Select map, satellite, and street views; Search for POIs; Search by voice	Google	http://www.google.com/mobile/navigation/
Nokia	Nokia N9; Nokia N97 mini; Nokia 6710 Navigator; Nokia X6 16GB; Nokia X6 32GB; Nokia 5800 XpressMusic; Nokia 5800 Navigation Edition; Nokia 5230; Nokia E52; Nokia E72	GPS-enabled navigation; Real-time traffic; Driver and pedestrian navigation; Pedestrian navigation through pedestrian networks for over 100 cities worldwide; Routes and directions via voice and text; Safety cameras and speed warnings; 3D landmarks	Ovi Maps	http://www.nokiausa.com/ovi-services-and-apps/ovi-maps/ovi-maps-main
Blackberry	BlackBerry® Storm™; BlackBerry® Curve™ 8330; BlackBerry® Pearl™ 8130; BlackBerry® 8830	Speech recognition; Traffic commute alerts; Proactively searches every five minutes for traffic congestion or incidents along drivers' routes; Current weather condition and forecast 7-days weather; Routes and directions via voice; 3D maps; Re-route if driver deviates from computed route; Re-route to avoid traffic; Include more than 10 million business listings such as gas prices and business reviews	TeleNav GPS Navigator	http://www.telenav.com/about/pr/pr-20090401.html

options based on nearby bus stops. It also allows users to select map, satellite, and street views, search for POIs, search by voice, and search along route.

Through third-party resources such as Ovi Maps, Nokia devices provide GPS-enabled navigation, real-time traffic, navigation for drivers and pedestrians; pedestrian navigation through pedestrian networks is supported for over 100 cities worldwide. Directions, which could be provided via voice or text, include safety cameras, speed warnings, and 3D landmarks.

Through third-party resources such as TeleNav GPS Navigator, Blackberry devices feature speech recognition and provide traffic commute alerts and proactively search every five minutes for traffic congestion or incidents along drivers' routes. They provide current weather condition and forecast 7-day weather, routes and directions via voice, 3D maps, and re-route if the driver deviates from the computed route and to avoid traffic. They also include more than 10 million business listings such as gas prices and business reviews.

4.5   Summary

In this chapter, the concept of UNAVIT as a platform that facilitates universal navigation is discussed. UNAVIT is intended to be universal with respect to personalized navigation assistance which can provide navigation anywhere, anytime, and for any user. The main features of UNAVIT are AnyWhere (navigation assistance in outdoors and indoors), AnyTime (navigation assistance suitable for different days, times, situations), AnyUser (navigation assistance to the general population and to the special needs population), AnyModeOfTravel (navigation assistance to drivers, pedestrians, bikers, wheelchair riders), Transparency (providing navigation assistance without user intervention), and Adaptability (adapting navigation assistance to user's needs and navigational behavior changes). An architecture of UNAVIT is presented and smartphones as the gateways to UNAVIT are discussed. Given that navigation services are expected to be scalable (increase in navigation database sizes and large simultaneous navigation users), cloud computing is considered as a suitable platform for UNAVIT. WMSs and some dominant WMS providers are disused. A summary of current navigation services provided by some navigation service providers is given.

Further Readings

Akasaka Y, Onisawa T (2008) Personalized Pedestrian Navigation System with Subjective Preference Based Route Selection. In: Intelligent Decision and Policy Making Support Systems(Ruan, D., ed), pp 73-91.

Bansal M, Jung SH, Matei B, Eledath J, Sawhney H (2010) Combining structure and appearance cues for real-time pedestrian detection. In: Proceeding on the International Society for Optical Engineering (SPIE), vol. 7692 Orlando, Florida, USA.

Bessho M, Kobayashi S, Koshizuka N, Sakamura K (2008) uNavi: Implementation and deployment of a place-based pedestrian navigation system. In: Proceedings of International Computer Software and Applications, pp 1254-1259.

Cho SY, Park CG (2006) MEMS based pedestrian navigation system. Journal of Navigation 59:135-153.

Feliz R, Zalama E, García-Bermejo JG (2009) Pedestrian tracking using inertial sensors. Journal of Physical Agents 3:35-42.

Flora Cd, Ficco M, Russo S, Vecchio V (2005) Indoor and Outdoor Location Based Services for Portable Wireless Devices. In: the 25th IEEE International Conference on Distributed Computing Systems Workshops (ICDCSW'05), pp 244-250 Columbus, OH, USA: IEEE.

Ghafourian M, Karimi H (2009) Universal Navigation: Concept and Algorithms. In: World Congress on Computer Science and Information Engineering (CSIE 2009), Anaheim, Los Angeles, March 31-April 2 2009. IEEE, pp 369-373.

Hile H, Vedantham R, Cuellar G, Liu A, Gelfand N, Grzeszczuk R, Borriello G (2008) Landmark-based pedestrian navigation from collections of geotagged photos. In: Proceedings of the 7th International Conference on Mobile and Ubiquitous Multimedia (MUM'08), pp 145-152.

Holone H (2007) Users are doing it for themselves: Pedestrian navigation with user generated content. In: Proceeding of the 2007 International Conference on Next Generation Mobile Applications, Services and Technologies (NGMAST 2007), pp 91-99.

Karimi H, A, Ghafourian, M. (2009) Universal Navigation. In: GIM International, pp 17-19.

Kourogi M, Sakata N, Okuma T, Kurata T (2006) Indoor/Outdoor Pedestrian Navigation with an Embedded GPS/RFID/Self-contained Sensor System. In: International Conference on Artificial Reality and Telexistence No16, vol. 4282, pp 1310-1321 Hangzhou, CHINE.

Lee HH, Park IK, Hong KS (2008) Design and implementation of a mobile devices-based real-time location tracking. In: The Second International Conference on Mobile Ubiquitous Computing, Systems, Services and Technologies (UBICOMM '08), pp 178-183 Valencia.

Li C (2006) User preferences, information transactions and location-based services: A study of urban pedestrian wayfinding. Computers, Environment and Urban Systems 30:726-740.

Mahler T, Reuff M, Weber M (2007) Pedestrian navigation system implications on visualization. In: Proceedings of the 4th international conference on Universal access in human-computer interaction: ambient interaction Beijing, China.

May AJ, Ross T, Bayer SH, Tarkiainen MJ (2003) Pedestrian navigation aids: information requirements and design implications. Personal and Ubiquitous Computing 7.

Park SK, Suh YS (2009) Gait state classification by hmms for pedestrian inertial navigation system. Transactions of the Korean Institute of Electrical Engineers 58:1010-1018.

Petrovski I, Okano K, Ishii M, Torimoto H, Konishi Y, Shibasaki R (2003) Pedestrian ITS in Japan: Pseudolites and GPS. GPS World 14:33-37.

Rehrl K, Göll N, Leitinger S, Bruntsch S (2005) Combined indoor/outdoor Smartphone navigation for public transport travelers. In: 3rd Symp LBS & TeleCartography(Gartner, G., ed), pp 235-239 Vienna, Austria.

Retscher G (2007) Test and integration of location sensors for a multi-sensor personal navigator. Journal of Navigation 60:107-117.

Retscher G, Kealy A (2006) Ubiquitous positioning technologies for modern intelligent navigation systems. Journal of Navigation 59:91-103.

Retscher G, Thienelt M (2004) NAVIO- A Navigation and Guidance Service for Pedestrians. Journal of Global Positioning Systems 3:208-217.

Santos AC, Tarrataca L, Cardoso JMP (2010) The Feasibility of Navigation Algorithms on Smartphones using J2ME. Mobile Networks and Applications 1-12.

Chapter 5
Anywhere Navigation

5.1 Introduction

Anywhere navigation in universal navigation refers to UNAVIT's ability that provides navigation assistance regardless of location, which could be outdoor (any geographic area) or indoor (any building). While today there are navigation systems/services that provide navigation assistance in outdoors or in indoors, there are no commercial navigation systems/services (though there are experimental systems/services) that allow seamless transition from outdoor to indoor or from indoor to outdoor. This means that in the absence of UNAVIT, the user seeking navigation assistance for moving from outdoor to indoor or vice versa must carry different equipments on them and manually switch from one to another as the environment changes. Switching from one environment to another requires changes of geo-positioning sensors and techniques, change of map databases, change of networks, change of map matching algorithms, among other changes. Obviously, this is impractical and expensive.

As one feature of UNAVIT, anywhere navigation allows transition between the two environments (i.e., outdoors and indoors) as seamless as possible and without user intervention. In this chapter, an ontology defining the concepts and the relationships among them for the anywhere feature are discussed and two algorithms, one for seamless transition from outdoor to indoor and one for seamless transition from indoor to outdoor, are presented.

5.2 Ontology

Figure 5.1 shows an ontology for the anywhere feature of UNAVIT, depicting situations when user's movement is from outdoor to indoor and from indoor to outdoor. To distinguish between the concepts associated with outdoor navigation and indoor navigation, the concepts in this ontology are grouped into activities related to navigation in outdoors and navigation in indoors and the relationships among them. As shown in this ontology, user's request for navigation would be either for navigation in outdoors or navigation in indoors.

H. A. Karimi, *Universal Navigation on Smartphones*,   89
DOI 10.1007/978-1-4419-7741-0_5, © Springer Science+Business Media, LLC 2011

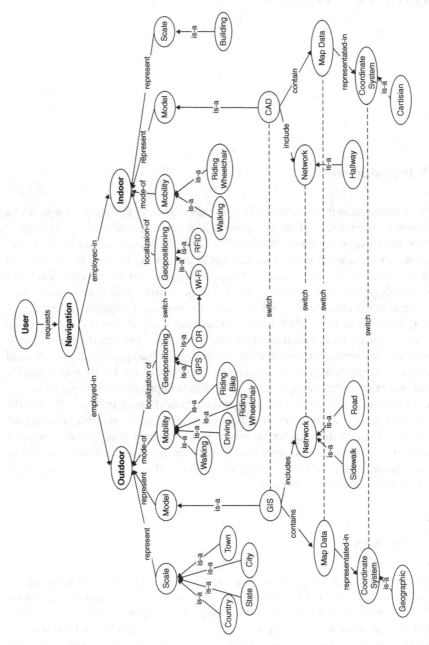

Fig. 5.1 Anywhere ontology

With respect to user's request for navigation assistance in outdoor, the concept of "outdoor" contains four sub-concepts: mobility, geo-positioning, model, and scale. The "mobility" sub-concept, which is related to the "outdoor" concept through the "mode-of" relation, is for different modes of travel a user may request for outdoor navigation which could be "walking", "driving", "riding wheelchair", or "riding bike", all through the "is-a" relation. The "geo-positioning" concept is related to the "outdoor' concept through the "localization-of" relation and is for different geo-positioning sensors which are possible for outdoor navigation including GPS and DR, each related to "geo-positioning" through the "is-a" relation. The "model" sub-concept that is related to the "outdoor" concept through the "represent" relation is for the type of data required for outdoor navigation which is dominantly "GIS" and is related to the "model" sub-concept through the "is-a" relation. The data in the "GIS" model includes "network" which, through the "is-a" relation, could be "road" or "sidewalk". The "GIS" model contains "map data" for navigation that can be represented in "coordinate system" through the "represented-in" relation which is a (through the "is-a" relation) "geographic" coordinate system. The "scale" sub-concept that is related to the "outdoor" concept through the "represent" relation is for the geographic extent within which outdoor navigation is requested and could be "country", "state", "city", or "town", all related to the "scale" sub-concept through the "is-a" relation.

With respect to user's request for navigation assistance in indoor, the concept of "indoor" contains four sub-concepts: mobility, geo-positioning, model, and scale. The "mobility" sub-concept which is related to the "indoor" concept through the "mode-of" relation is for different modes of travel a user may request for indoor navigation which could be "walking" or "riding wheelchair", both through the "is-a" relation. The "geo-positioning" sub-concept which is related to the "indoor" concept through the "localization-of" relation is for different geo-positioning sensors that are possible in indoor navigation including RFID and Wi-Fi, each related to "geo-positioning" through the "is-a" relation. The "model" sub-concept, which is related to the "indoor" concept is for the type of data required for indoor navigation which is dominantly "CAD" and is related to the "model" sub-concept through the "is-a" relation. The data in the "CAD" model includes "network" which, through the "is-a" relation is "hallway". The "CAD" model contains "map data" for navigation that can be represented in "coordinate system" through the "represented-in" relation, which is a (through the "is-a" relation) "Cartesian" coordinate system. The "scale" sub-concept that is related to the "indoor" concept through the "represent" relation is for the geographic extent within which indoor navigation is requested and can be "building" related to the "scale" sub-concept through the "is-a" relation.

The ontology also represents knowledge about the sub-concepts in the "outdoor" and "indoor" concepts that must be switched when transition from one environment to the other is considered; the dashed lines in the diagram (Figure 5.1) show the sub-concepts in the corresponding environments that must be switched. These sub-concepts are "geo-positioning" in both outdoor and indoor navigation and "GIS" in outdoor navigation and "CAD" in indoor navigation. As shown in the ontology, once transition from one environment to the other is detected, "navigation data", "network", and "coordinate system" must be switched.

5.3 Analysis

Anywhere navigation may be grouped into five categories: indoor navigation, outdoor navigation, indoor outdoor navigation, outdoor-indoor navigation, and indoor-outdoor indoor navigation. Each of these categories is described below.

Indoor navigation is the instance where navigation assistance is required in indoors only (i.e., origin and destination locations and the entire route are within the same building). Indoor navigation can be expressed as:

$$\text{Route} = R_i = \{h_1, h_2, \ldots, h_n\}$$

$$\text{Distance} = \sum_{i=1}^{n} d\,(h_i)$$

$$\text{Compute} = C_i$$

$$\text{Direction} = D_i$$

where Route R_i is the set of hallway segments, h_i, forming the computed route in the building; Distance is the total length of the route calculated by accumulation of $d(h_i)$, length of each segment of the route; Compute is the type of computation, C_i, which is route computation in the building; Direction is the set of directions, D_i, in the building.

Outdoor navigation is the instance where navigation assistance is required in outdoor only (i.e., both origin and destination locations and the entire route all are in the outdoor). Outdoor navigation can be expressed as:

$$\text{Route} = R_j = \{s_1, s_2, \ldots, s_m\}$$

$$\text{Distance} = \sum_{j=1}^{m} d\,(s_j)$$

$$\text{Compute} = C_o$$

$$\text{Direction} = D_o$$

where Route R_j is the set of road/sidewalk segments, s_j, forming the computed route in the outdoor; Distance is the total length of the route calculated by accumulation of $d(s_j)$, length of each segment of the route; Compute is the type of computation, C_o, which is route computation in the outdoor; Direction is the set of directions, D_o, in the outdoor.

Indoor-outdoor navigation is the instance where navigation assistance is required both in indoors and outdoors, i.e., origin location is in the indoor, destination location is in the outdoor, and the route is partly in the indoor and partly in the outdoor. Indoor-outdoor navigation can be expressed as follows:

$$Route = R_i \cup R_j = \{h_1, h_2, \ldots, h_n\} \cup \{s_1, s_2, \ldots, s_m\}$$

$$Distance = \sum_{i=1}^{n} d\,(h_i) + \sum_{j=1}^{m} d\,(s_j)$$

$$Compute = C_i \rightarrow C_o$$

$$Direction = D_i \rightarrow D_o$$

where Route $R_i \cup R_j$ is the union of the set of hallway segments in the indoor, h_i, and the set of road/sidewalk segments in the outdoor, s_j, forming the computed route from indoor to outdoor; Distance is the total length of the route by calculated by accumulation of $d(h_i)$, length of each segment of the portion of the route in the indoor, and accumulation of $d(s_j)$, length of each segment of the portion of the route in the outdoor; Compute is the type of computation, C_i, which is route computation in the indoor, followed by C_o, which is route computation in the outdoor; Direction is the set of directions, D_i, in the indoor, followed by the set of directions, D_o, in the outdoor.

Outdoor-indoor navigation is the instance where navigation assistance is required both in outdoors and indoors (i.e., origin location is in the outdoor, destination location is in the indoor), and the route is partially in the outdoor and partially in the indoor. Outdoor-indoor navigation can be expressed as follows:

$$Route = R_j \cup R_i = \{s_1, s_2, \ldots, s_m\} \cup \{h_1, h_2, \ldots, h_n\}$$

$$Distance = \sum_{j=1}^{m} d\,(s_j) + \sum_{i=1}^{n} d\,(h_i)$$

$$Compute = C_o \rightarrow C_i$$

$$Direction = D_o \rightarrow D_i$$

where Route $R_j \cup R_i$ is the union of the set of road/sidewalk segments in the outdoor, s_j, and the set of hallway segments in the indoor, h_i, forming the computed route from outdoor to indoor; Distance is the total length of the route calculated by accumulation of $d(s_j)$, length of each segment of the portion of the route in the outdoor, and accumulation of $d(h_i)$, length of each segment of the portion of the route in the outdoor; Compute is the type of computation, C_o, which is route computation in the outdoor, followed by C_i, which is route computation in the indoor; Direction is the set of directions, D_o, in the outdoor, followed by the set of directions, D_i, in the indoor.

Indoor-outdoor-indoor navigation is the instance where navigation assistance is required both in outdoors and indoors where origin location is in the indoor, destination location is also in the indoor (different than the indoor where origin is located), and the route is partially in the indoors and partially in the outdoor. Indoor-outdoor-indoor navigation can be expressed as follows:

$$\text{Route} = R_i \cup R_j \cup R_k = \{h_1, h_2, \ldots, h_n\} \cup \{s_1, s_2, \ldots, s_m\} \cup \{h_1, h_2, \ldots, h_k\}$$

$$\text{Distance} = \sum_{i=1}^{n} d\,(h_i) + \sum_{j=1}^{m} d\,(s_j) + \sum_{i=1}^{p} d\,(h_k)$$

$$\text{Compute} = (H_{i1}, C_{i1}) \rightarrow (R_o, C_o) \rightarrow (H_{i2}, C_{i2})$$

$$\text{Direction} = (H_{i1}, D_{i1}) \rightarrow (R_o, D_o) \rightarrow (H_{i2}, D_{i2})$$

where Route $R_i \cup R_j \cup R_k$ is the union of the set of hallway segments in the first building, h_i, the set of road/sidewalk segments in the outdoor, s_j, and the set of hallway segments in the second building, h_k, forming the computed route from indoor to outdoor and from outdoor to indoor; Distance is the total length of the route calculated by accumulation of $d(h_i)$, length of each segment of the portion of the route in the first building, accumulation of $d(s_j)$, length of each segment of the portion of the route in the outdoor, and accumulation of $d(s_k)$, length of each segment of the portion of the route in the second building; Compute is the type of computation, (H_{i1}, C_{i1}), which is route computation in the first building, followed by (R_o, C_o), which is route computation in the outdoor, followed by (H_{i2}, C_{i2}), which is route computation in the second building; Direction is the set of directions, (H_{i1}, D_{i1}), in the first building, followed by the set of directions, (R_o, D_o) in the outdoor, followed by the set of directions, (H_{i2}, D_{i2}), in the second building.

5.4   Algorithms

In this section, two algorithms, one for transition from outdoor to indoor and another for transition from indoor to outdoor, are discussed. These algorithms address most common cases when a user moves from one environment to another. While the number of possibilities could be quite large, for illustrative purposes Figure 5.2 shows three cases. In these cases, the following assumptions are

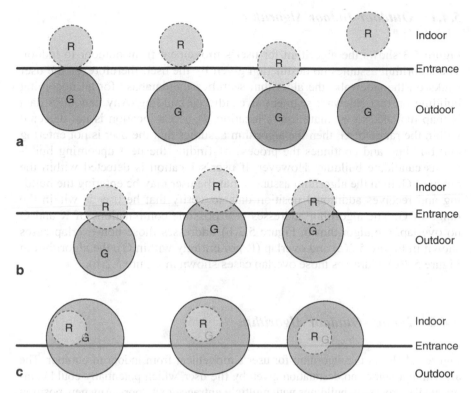

Fig. 5.2 Different cases between G (GPS in outdoor) and R (RFID in indoor). **a** R and G do not overlap (R ∩ G = ∅). **b** R and G overlap but R is not entirely within G (R ∩ G ≠ ∅ and R ⊄ G). **c** R and G overlap but R is entirely within G (R ⊂ G)

made: (a) user's navigation device is equipped with GPS for outdoor navigation and is able to detect RFID tags for indoor navigation; (b) there is an RFID tag near the entrance/exit door; (c) both GPS and RFID have a fixed but different error range (these are shown in Figure 5.2); G represents the error range of GPS and R represents the error range of RFID; and (d) R, the error range of RFID, is smaller than G, the error range of GPS.

Figure 5.2(a) shows the cases where the GPS error range and the RFID error range do not overlap. Figure 5.2(b) shows the cases where the GPS error range and the RFID error range overlap and R is not entirely within G. Figure 5.2(c) shows the cases where R is entirely within G.

The two algorithms are designed to handle these cases above and be able to detect three scenarios with respect to the movement of the user: (a) user's movement indicates a potential transition between the two environments; (b) the time when transition between the two environments occurs; and (c) the location where the transition between the two environments occurs.

5.4.1   Outdoor-Indoor Algorithm

Figure 5.3 shows the algorithm for user's movement from outdoor to indoor. The algorithm assumes no destination given by the user, therefore, as the user walks on the sidewalk, the algorithm searches the database for an upcoming building on that sidewalk segment as candidate building. Any new GPS data is map matched to estimate user's location. If user's location is not detected within the range of G, then the algorithm assumes that the user is not entering that building and continues the process of finding the next upcoming building as candidate building. However, if user's location is detected within the range of G, then the algorithm assumes that the user may be entering the building and receives additional position data to verify that he user is within the range of G. The algorithm addresses two possible combinations of R and G no overlap, the algorithm in Figure 5.3(b) addresses those non-overlap cases shown in Figure 5.2(a) and overlap (R not entirely within G), the algorithm in Figure 5.3(c) addresses those overlap cases shown in Figure 5.2(b).

5.4.2   Indoor-Outdoor Algorithm

Figure 5.4 shows the algorithm for user's movement from indoor to outdoor. The algorithm assumes no destination given by the user, which potentially could complicate the process in buildings with multiple entrance/exit doors. Any new position data by RFID is map matched to estimate user's location. As the user walks on the hallway of a building, the algorithm checks for RFID signals near the entrance/exit door. If user's location is not detected within the range of R, then the algorithm assumes that the user is not exiting the building and continues the process of finding user's location within the building. If user's location is detected within the range of R then the algorithm assumes that the user may be exiting the building and receives additional position data to verify that the user is within the range of R. The algorithm addresses two possible combinations of R and G no overlap, the algorithm in Figure 5.4(b) addresses those non-overlap cases shown in Figure 5.2(a) and overlap (R not entirely within G), the algorithm in Figure 5.4(c) addresses those cases shown in Figure 5.2(b).

Figure 5.5[1] shows a route from an origin in indoor to a destination in outdoor. The total route has two parts, one in indoor and one in outdoor. The indoor part of the route uses the hallway network of the building and a set of user's preferences to compute a route from the origin to the entrance/exit door. The outdoor part of the route uses the sidewalk network in the area connected to the building and a different set of user's preferences to compute a route from the entrance/exit door of the building to the destination. Even when both sets of user's preferences for routes in

---

[1] http://nees.berkeley.edu/Facilities/images/building484a.jpg

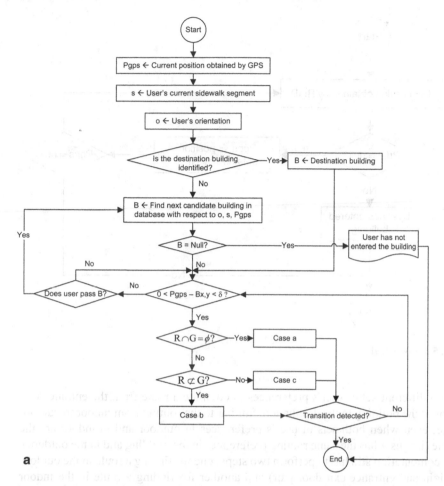

Fig. 5.3 Outdoor-indoor algorithm cases. **a** Outdoor-indoor algorithm. **b** Outdoor-indoor algorithm (non-overlapping case). **c** Outdoor-indoor algorithm for overlap but R not entirely within G case

indoor and in outdoor are the same (i.e., the user has the same routing preferences in the building and in the outdoor), the computation still has to perform two steps: one for finding a route in the indoor (origin and entrance/exit door pair) and another for finding a route in the outdoor (entrance/exit door and destination pair).

Figure 5.6[2] shows a route from an origin in outdoor to a destination in indoor. The total route has two parts, one in outdoor and one in indoor. The outdoor part of the route uses the sidewalk network of the area connected to the building and a set of user's preferences to compute a route from the origin to the entrance/exit door of the building. The indoor part of the route uses the hallway network of the building

---

[2] http://rufrealty.com/Collegeblvd/bldglyout/images/CBPOffice2.gif

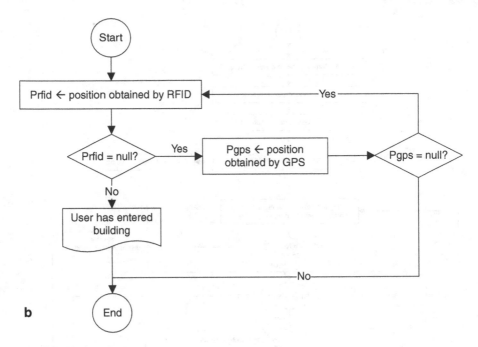

b

Fig. 5.3 Continued

and a different set of user's preferences to compute a route from the entrance/exit door of the building to the destination. Similar to the routing from indoor to outdoor case, even when both sets of user's preferences in outdoor and in indoor are the same (i.e., user has the same routing preferences in the building and in the outdoor), the computation still has to perform two steps: one for finding a route in the outdoor (origin and entrance/exit door pair) and another for finding a route in the indoor (entrance/exit door and destination pair).

Figure 5.7[3] shows a route from an origin in indoor, passing through outdoor, to a destination in indoor. The total route has three parts, the first part in indoor, the second part in outdoor, and the third part in indoor. The first part of the route in indoor uses the hallway network of the building where the origin is located and a set of user's preferences to compute a route from the origin to the entrance/exit door of the building. The second part of the route in outdoor uses the sidewalk network in the area connected to the building where the origin is located and a different set of user's preferences to compute a route from the entrance/exit door of the building to the entrance/exit door of the building where the destination is located. The third part of the route in indoor uses the hallway network of the building where the destination is located and a different set of user's preferences to compute a route from the entrance/exit door of the building to

---

[3] http://bcs.solano.edu/building.JPG

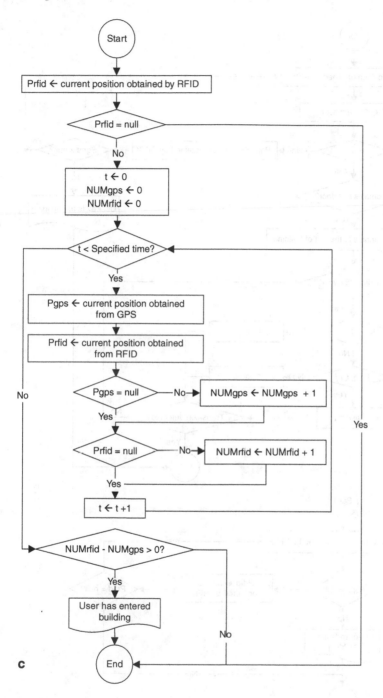

Fig. 5.3 Continued

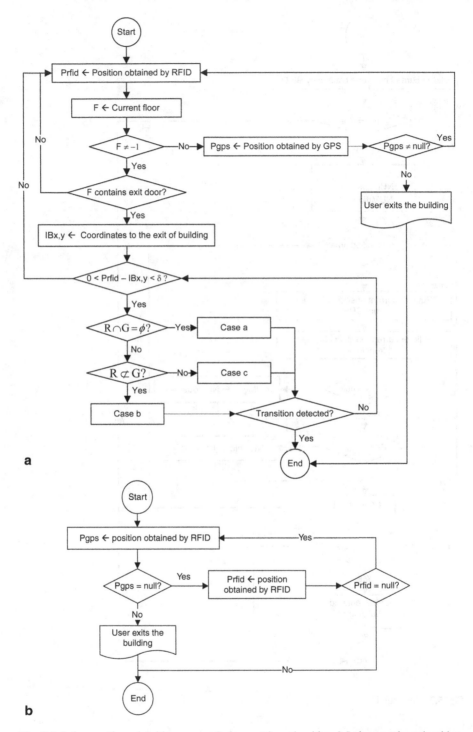

Fig. 5.4 Indoor-outdoor algorithm cases. **a** Indoor-outdoor algorithm. **b** Indoor-outdoor algorithm (non-overlapping case). **c** Indoor-outdoor algorithm for overlap but R not entirely within G case

Fig. 5.4 Continued

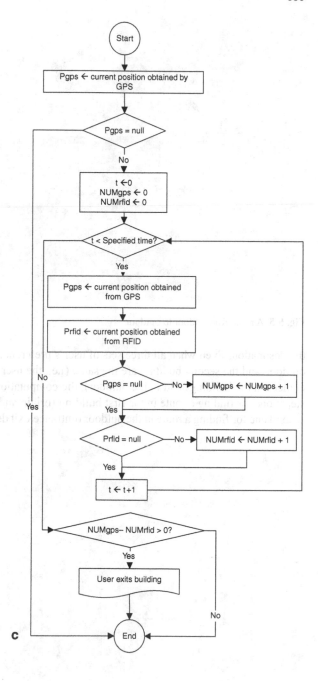

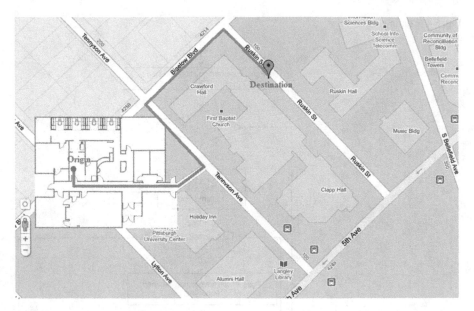

Fig. 5.5 A route from indoor to outdoor

the destination. Even when all three sets of user's preferences for the first building, the outdoor, and the second building are the same (i.e., the user has the same routing preferences in the buildings and in the outdoor), the computation still has to perform three steps: one for finding a route in the first building (origin and entrance/exit door pair), a second one for finding a route in the outdoor (entrance/exit door of the first building and

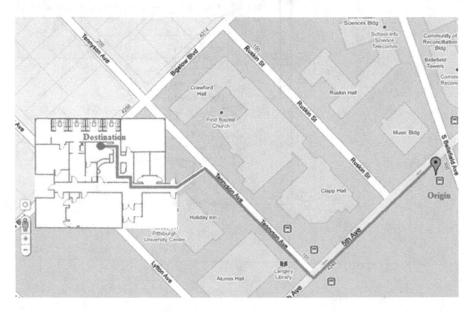

Fig. 5.6 A route from outdoor to indoor

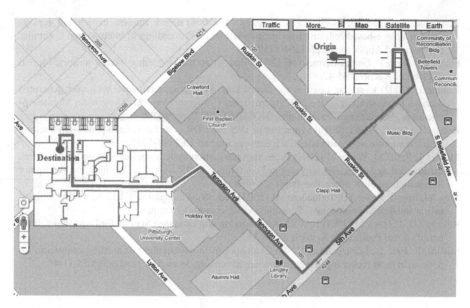

Fig. 5.7 A route from indoor to outdoor to indoor

entrance/exit door of the second building pair), and the third one for finding a route in the second building (entrance/exit door and destination pair).

5.5 Summary

In this chapter, the anywhere feature of UNAVIT is discussed. Anywhere is defined as the ability of UNAVIT for providing navigation assistance regardless of user's location, indoor or outdoor. Possible cases for navigation assistance are indoor navigation, outdoor navigation, indoor-outdoor navigation, outdoor-indoor navigation, and indoor-outdoor-indoor navigation. Two algorithms, one for transition from indoor to outdoor and one for transition from outdoor to indoor, are discussed and analyzed. Although these two algorithms are based on GPS and RFID as the geo-positioning sensors for determining user's location in outdoors and indoors, respectively, they can easily be adjusted for other geo-positioning sensors.

Further Readings

Akasaka Y, Onisawa T (2008) Personalized Pedestrian Navigation System with Subjective Preference Based Route Selection. In: Intelligent Decision and Policy Making Support Systems (Ruan, D., ed), pp 73-91.

Bessho M, Kobayashi S, Koshizuka N, Sakamura K (2008) uNavi: Implementation and deployment of a place-based pedestrian navigation system. In: Proceedings of International Computer Software and Applications, pp 1254-1259.

Feliz R, Zalama E, García-Bermejo JG (2009) Pedestrian tracking using inertial sensors. Journal of Physical Agents 3:35-42.

Holone H (2007) Users are doing it for themselves: Pedestrian navigation with user generated content. In: Proceeding of the 2007 International Conference on Next Generation Mobile Applications, Services and Technologies (NGMAST 2007), pp 91-99.

Huang C, Liao Z, Zhao L (2010) Synergism of INS and PDR in self-contained pedestrian tracking with a miniature sensor module. IEEE Sensors Journal 10:1349-1359.

Mahler T, Reuff M, Weber M (2007) Pedestrian navigation system implications on visualization. In: Proceedings of the 4th international conference on Universal access in human-computer interaction: ambient interaction Beijing, China.

May AJ, Ross T, Bayer SH, Tarkiainen MJ (2003) Pedestrian navigation aids: information requirements and design implications. Personal and Ubiquitous Computing 7.

Retscher G, Thienelt M (2004) NAVIO- A Navigation and Guidance Service for Pedestrians. Journal of Global Positioning Systems 3:208-217.

Stirling R, Fyfe K, Lachapelle G (2005) Evaluation of a new method of heading estimation for pedestrian dead reckoning using shoe mounted sensors. Journal of Navigation 58:31-45.

Chapter 6
Anytime Navigation

6.1 Introduction

Anytime navigation, a capability of universal navigation, refers to UNAVIT's ability that provides navigation assistance based on the requirements and constraints of different times in a day, different days in a week, different months, and different seasons given specific locations. In other words, UNAVIT is capable of providing navigation assistance based on both spatial and temporal variations. For example, for an origin-destination pair, the least congested route during rush hours and the safest route during the night hours may be considered.

Figure 6.1 shows time and situation as two factors influencing navigation in outdoors. For example, a route optimal for navigation during rush hours is different than the one which is optimal for navigation during non-rush hours. Similarly, a route optimal for navigation on a dry road is different than the one which is optimal for navigation on a wet road.

Anytime feature of universal navigation mainly concerns outdoor navigation, based on such factors and criteria as summarized in Table 6.1. The table shows that navigation could be influenced by time of day, day of week, and month and season. Depending on a time of travel and the factor(s) influencing navigation based on a specific mode of travel, UNAVIT will take into account different criteria for navigation and routing.

Factors influencing navigation during a day include non-rush hours, rush hours, daylight, and dark. Factors influencing navigation in a week include weekdays, weekends, holidays, and events. Factors influencing navigation in a season include weather, dry condition and wet condition. Each time and mode of travel (i.e., driving, biking, and walking) requires consideration of a different set of navigation criteria. Such criteria include congestion, safety, and accident.

Another important consideration in anytime navigation is purpose of trip, which could be commute, leisure, or emergency. This means that the factors and criteria in Table 6.1 may be modified depending on purpose of trip. For example, when purpose of trip is commute, rush hours may be a factor, and when purpose of trip is leisure, driving through scenic areas may be preferred.

H. A. Karimi, *Universal Navigation on Smartphones,*
DOI 10.1007/978-1-4419-7741-0_6, © Springer Science+Business Media, LLC 2011

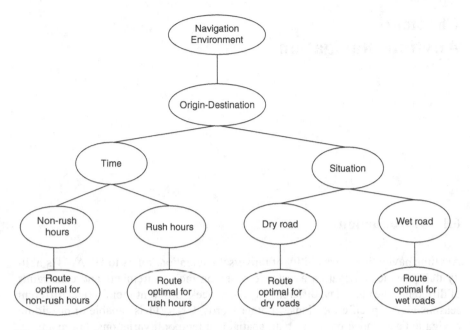

Fig. 6.1 Factors influencing navigation in outdoors

Unlike navigation in outdoors, where factors at different times may influence navigation activities, time has no impact on navigation activities in indoors. For example, change of weather, rain or sun, does not impact the way one would walk in a building. The only concern, however, in indoor navigation is accessibility at different times. For public buildings, different times that could impact accessibility are weekdays, weekends, and holidays, where during certain hours the entire building, certain floors, or certain rooms may not be accessible to all or some users.

Table 6.1 Time factors influencing navigation

Time	Factors	Mode of Travel (Navigation Criteria)		
		Driving	Biking	Walking
Day	Regular hours	Congestion	Congestion	Safety
	Rush hours	Safety	Safety	
	Day			
	Night			
Week	Weekdays	Congestion	Congestion	Safety
	Weekends			
	Holidays			
	Events			
Season	Weather	Congestion	Congestion	Safety
	Dry condition	Slippage	Slippage	
	Wet condition (rain, snow)	Accident	Accident	

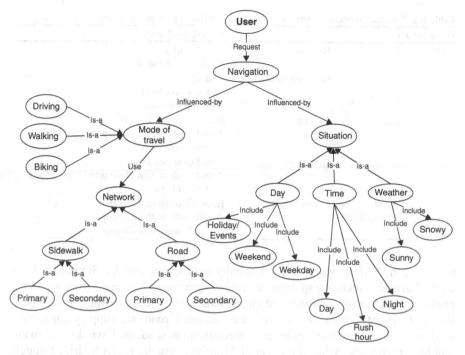

Fig. 6.2 An ontology for anytime navigation

6.2 Ontology

Figure 6.2 shows an ontology for anytime navigation. The ontology represents various concepts including time, situation, mode of travel, and type of network. In this ontology, situation is a high-level concept that contains three sub-concepts: day, time, and weather. Day may include weekday, weekend holiday and event. Time may include day, night, and rush hour. Weather could be sunny or snowy. The mode of travel concept, which could be driving, walking, or biking, uses different networks. The networks for navigation in outdoors could be road networks or sidewalk networks, where each network segment (road or sidewalk) is either of primary or secondary category.

6.3 Requirements

Table 6.2 shows time, event, and weather condition when navigation needs of drivers may change. The table shows that drivers during non-rush hours typically prefer shortest and fastest routes and prefer to avoid main roads and congested roads during rush hours. During holidays, drivers usually would prefer to avoid congested

Table 6.2 Navigation needs of drivers in outdoors in different times/situations

Time/Situation			Navigation Needs
Day	Time	Rush hours	Avoid main roads
			Avoid congested roads
		Non-rush hours	Usually:
			Shortest distance route
			Fastest time route
	Event	Holidays	Avoid congested roads
			Avoid roads within close proximity to shopping centers
		Game/Show/Concert	Avoid congested roads
			Avoid roads within close proximity to location of the event
	Weather condition	Snowy/Rainy	Route through main roads
			Avoid roads with high slope
Night			Route through well-lit roads

roads and those roads within close proximity to shopping centers. Similarly, in the case of an event, such as a sport game, a show, or a concert, drivers would typically prefer to avoid congested roads and those roads within close proximity to the location of the event. Depending on weather condition, if roads are slippery due to rain or snow, drivers would prefer to drive through main roads and avoid those roads that have high slope. When driving at night, drivers usually prefer to drive through well-lit roads.

Table 6.3 shows time, event, and weather condition when navigation needs of pedestrians may change. The table shows that pedestrians during non-rush hours typically prefer shortest routes and prefer to avoid sidewalks through main roads during rush hours. During holidays, pedestrians usually would prefer to avoid congested sidewalks and those sidewalks within close proximity to shopping centers. Similarly, in the case of an event, such as a sport game, a show, or a concert, pedestrians would typically prefer to avoid congested sidewalks and those sidewalks

Table 6.3 Navigation needs of pedestrians in outdoors in different times/situations

Time/Situation			Navigation Needs
Day	Time	Rush hours	Avoid sidewalks through main roads
			Avoid congested sidewalks
		Non-rush hours	Usually:
			Shortest distance route
	Event	Holidays	Avoid congested sidewalks
			Avoid sidewalks within close proximity to shopping centers
		Game/Show/Concert	Avoid congested sidewalks
			Avoid sidewalks within close proximity to the location of the event
	Weather condition	Snowy/Rainy	Route through sidewalks on main roads
			Avoid sidewalks with high slope
Night			Route through well-lit sidewalks

Table 6.4 Navigation needs of pedestrians and wheelchair riders in indoors in different times/situations

Time/Situation			Navigation Needs
Day/Night	Time	Rush hours (e.g., before class time)	Avoid congested hallways
		Non-rush hours	Usually: Shortest route
	Event	Class/Seminar/Talk	Avoid hallways within close proximity to the location of the event

within close proximity to the location of the event. Depending on weather condition, if sidewalks are slippery due to rain or snow, pedestrians would prefer to walk on sidewalks of main roads and avoid those sidewalks that have high slope. When walking at night, pedestrians usually prefer to walk through well-lit sidewalks.

Unlike navigation in outdoors, which could adversely be impacted by environmental factors such as change of time, event, or weather condition, navigation in indoors is rarely impacted by environmental factors. However, it is conceivable that change of time and occurrence of events may cause interruption or delay in navigation in indoors. Table 6.4 shows time and event as two factors that may require change in navigation needs of pedestrians or wheelchair riders. As shown in this table, pedestrians or wheelchair riders during non-rush hours typically prefer shortest routes and prefer to avoid congested hallways during rush hours (e.g., before a class time in a school). In the case of an event, such as a class, a seminar, or a talk, pedestrians and wheelchair riders would typically prefer to avoid congested hallways within close proximity to the location of the event.

6.4   Algorithm

In general, there are two factors that may impose computation of new routes in UNAVIT, as shown in Figure 6.3. One factor is change of time which may lead to change of navigation needs. An example is a route that is optimal during daylight but may not be optimal at dark. Another example is a route that offers fastest travel time during the summer months when most schools are closed but is not fastest route during the school months, typically in fall and spring terms, when schools are open. An example event that may require a different route than shortest distance is a football game where the roads near the location of the stadium are expected to be congested during a period of time.

Routes can be computed based on three general situations at a given location: fixed-time, new-event, and new-time-new-event. In the fixed-time situation both location and time are fixed. In this situation, a route R_i at time t_i for a pair of origin and destination locations can be computed and expressed as

$$R(O - D, t_i) \rightarrow R_i$$

Fig. 6.3 Factors impacting
change of routes

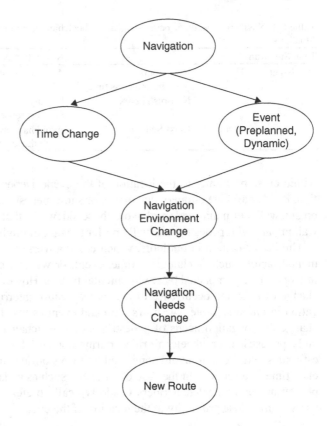

In the new-event situation computation of a new route upon occurrence of a new event is required. In this situation, a new route R_j based on an event (e.g., construction) for the same pair of origin and destination locations can be computed and expressed as

$$R(O - D, e_i) \rightarrow R_j$$

In the new-time-new-event situation computation of a new route upon change of time and occurrence of a new event is required. In this situation, a new route R_{ij} based on an event (e.g., accident) at time t_i for the same pair of origin and destination locations can be computed and expressed as

$$R(O - D, t_i, e_j) \rightarrow R_{ij}$$

Figure 6.4 shows the algorithm for handling anytime navigation in UNAVIT. The anytime algorithm aims at realizing the navigation requirements at different times and in different situations in order to provide assistance that is most suitable for the given location and time. The algorithm starts by realizing mode of travel, which could be driving, biking, or walking.

If mode of travel is driving, the algorithm checks for time of navigation. If navigation is during rush hours, the algorithm tries to avoid congested roads. Similarly,

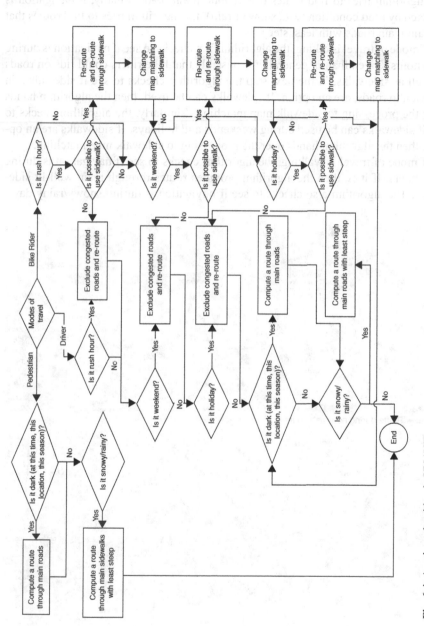

Fig. 6.4 Anytime algorithm in UNAVIT

if navigation is during weekends and holidays, the algorithm tries to avoid congested roads. In the case where navigation is during night hours, when it is dark, the algorithm tries to find routes that contain main roads. Finally, if navigation is affected by road condition (e.g., snow or rain), the algorithm tries to find routes that contain main roads with least steep.

If mode of travel is biking, the algorithm first checks to see if navigation is during rush hours. Given that, in general, it is possible that bike riders could ride on roads as well as sidewalks, in the next step the algorithm checks to see if sidewalks, in addition to roads, are an option. If sidewalks can be used, then the algorithm hands over the processing to sidewalk map matching. Similarly, the algorithm checks to see if sidewalks can be used during weekends and holidays. If sidewalks are an option, then the algorithm hands over the processing to sidewalk map matching.

If mode of travel is walking, the algorithm checks to see if navigation is during dark hours. If it is, then the algorithm provides routes through main well-lit sidewalks. The algorithm also checks to see if navigation is during snowy/rainy days,

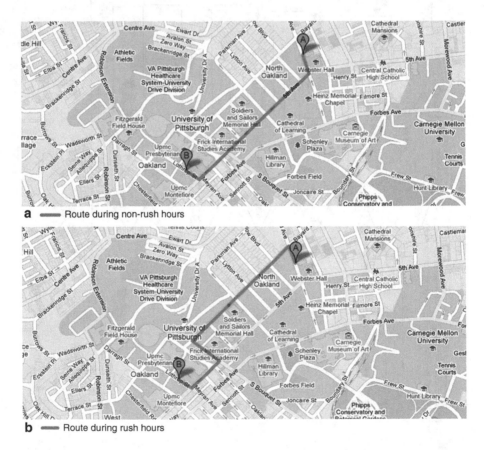

Fig. 6.5 Routes during different times. **a** An optimal route between A and B during non-rush hours. **b** An optimal route between A and B during rush hours

as usually wet road/sidewalk conditions cause falling and slipping. If it is, the algorithm provides routes through main sidewalks with least steep reducing chances of falling while walking.

Figure 6.5 shows two different optimal routes for the same pair of origin-destination pair. Figure 6.5(a) shows an optimal route suitable during non-rush hours and Figure 6.5(b) shows an optimal route suitable during rush hours.

6.5   Summary

In this chapter, the anytime feature of UNAVIT is discussed. Anytime is defined as the ability of UNAVIT for providing navigation assistance based on user's navigation needs and preferences at different times of a day, different days of a week (weekdays, weekends), and different months of a season (dry roads, wet/snowy roads). The anytime feature is mainly of concern in outdoor navigation as such factors as weather condition and road congestion impact navigation needs and preferences. The anytime feature is of little concern in indoor navigation as there are really no factors such as those in outdoors that would impact navigation needs and preferences. An algorithm that addresses time-dependent navigation in UNAVIT by taking into account different times and events and different modes of travel is presented.

Further Readings

Bent R, Van Hentenryck P (2004) A two-stage hybrid local search for the vehicle routing problem with time windows. Transportation Science 38 (4):515-530.

Braysy O, Gendreau M (2005) Vehicle routing problem with time windows, Part I: Route construction and local search algorithms. Transportation Science 39 (1):104-118.

Cordeau J, Laporte G, Mercier A (2001) A unified tabu search heuristic for vehicle routing problems with time windows. Journal of the Operational Research Society 52 (8):928-936.

Das Gupta C (2010) Application of GPS and infrared for car navigation in foggy condition to avoid accident. In: 2nd International Conference on Computer Engineering and Applications (ICCEA 2010), vol. 2, pp 238-241 Bali Island, Indonesia.

Goel A (2009) Vehicle Scheduling and Routing with Drivers' Working Hours. Transportation Science 43 (1):17-26.

Hashimoto H, Ibaraki T, Imahori S, Yagiura M (2006) The vehicle routing problem with flexible time windows and traveling times. Discrete Applied Mathematics 154 (16):2271-2290

Ichoua S, Gendreau M, Potvin J (2003) Vehicle dispatching with time-dependent travel times. European journal of operational research 144 (2):379-396.

Ioannou G, Kritikos M, Prastacos G (2001) A greedy look-ahead heuristic for the vehicle routing problem with time windows. Journal of the Operational Research Society 52 (5):523-537.

Jain S, Fall K, Patra R Routing in a delay tolerant network. In: Proceedings of the 2004 conference on Applications, technologies, architectures, and protocols for computer communications (SIGCOMM '04), Portland,Oregon, USA, 30 August - 3 September 2004. ACM, pp 145-158.

Mester D, Bräysy O (2005) Active guided evolution strategies for large-scale vehicle routing problems with time windows. Computers & Operations Research 32 (6):1593-1614.

Chapter 7
Anyuser Navigation

7.1 Introduction

Anyuser navigation, a capability of universal navigation, refers to UNAVIT's ability that provides navigation assistance to any user, with and without special needs and preferences. In other words, the anyuser feature of UNAVIT is a reference to personalized navigation assistance since it addresses the navigation needs and preferences at individual level. Figure 7.1 shows "general population" and "special needs" as two categories of navigation users. In this figure, general population is a reference to users with general mobility needs; in other words, users under the general population category require no special navigation assistance. Typically, users under the general population category may have navigation preferences for navigation assistance, primarily for routes. An example of navigation preferences is a route that has two parts, one for business which may be computed based on the fastest travel time criterion followed by another part for leisure which may be computed based on the most scenic criterion. In Figure 7.1, special needs is a reference to those users faced with mobility challenges; in other words, users under the special needs category require special navigation assistance. Navigation assistance for users with special needs can generally be divided into four groups: mobility impaired, visually impaired, cognitively impaired, and elderly. Like users under the general population category, individuals belonging to each group under the special needs category may have their own navigation preferences. Examples of preferences by the special needs groups are: a mobility impaired who is wheelchair-bound may prefer a route that has no inclined surface throughout (note that a small level of surface inclination is possible for wheelchair bound individuals); a visually impaired may prefer a route that has least number of steps; a cognitively impaired (e.g., someone with Alzheimer's) may prefer a route that has least number of turns; and an elderly may prefer a route that has both least number of turns and least number of steps.

Figure 7.2 illustrates the relationships between the two categories of users, general population and special needs, and between the different groups in the special needs category. This figure shows specifically that the physical environment, indoor or outdoor, has more possibilities for the users in the general population category compared to the users in the special needs category. From a computational point

H. A. Karimi, *Universal Navigation on Smartphones*,
DOI 10.1007/978-1-4419-7741-0_7, © Springer Science+Business Media, LLC 2011

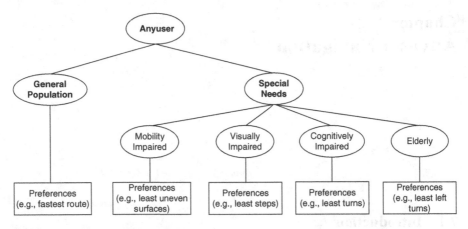

Fig. 7.1 General categories of users

of view this means that the solution space for the general population category is larger than the solution space for the special needs category. It is worth mentioning that current navigation systems/services mostly provide navigation assistance for users in the general population category. Figure 7.2 also shows that while each group in the special needs category has its own set of navigation requirements and preferences, there are overlapping requirements and preferences among the different groups. An example of a navigation requirement by all different groups in the special needs category is a route with low inclination. It should be noted that, as shown in Figure 7.2, the elderly group's navigation requirements and preferences overlap those of the other groups'.

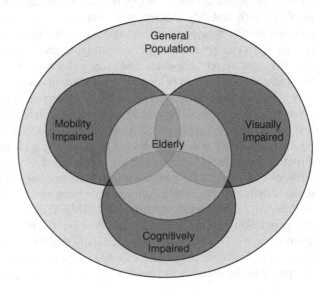

Fig. 7.2 Solution space for different user categories

7.2 Ontology

Figure 7.3 shows an ontology for the anyuser feature of UNAVIT. The ontology contains concepts in both categories, general population and special needs, and related to different navigation environments, indoor and outdoor. The sub-concept for navigation in outdoors by the general population includes different modes of travel: driving, riding, and walking. Each mode of travel includes routing criteria and there are common criteria in all modes of travel (e.g., shortest path). The sub-concept for navigation in indoors by the general population includes walking as the only mode of travel and the common criterion for routing in indoors is shortest path.

Figure 7.3 also shows the concepts related to the special needs category. These concepts include the group sub-concepts: mobility impaired, visually impaired, cognitively impaired, and elderly. The sub-concepts in each of these groups are navigation concepts in indoors and outdoors where each concept contains sub-concepts related to mode of travel.

Driving, riding wheelchair, or walking with walker are possible modes of travel by mobility impaired for navigation in outdoors. Routing preferences when driving is mode of travel by mobility impaired share the same preferences as the users in the general population category (shortest path, fastest time, least intersections). However, while riding wheelchair or walking with a walker, the mobility impaired would prefer to avoid steps, obstacles, slopes more than a certain amount, and curbs. In indoor, the possible modes of travel by mobility impaired are riding wheelchair and walking with walker, in which case routing preferences include no narrow hallways, no steps, and no obstacles.

The only possible mode of travel by visually impaired is walking, whether the navigation environment is outdoor or indoor. Possible routing preferences by visually impaired when walking outdoors are least obstacles, no unfamiliar areas, and no unsafe intersections. The major routing preference by visually impaired when walking indoors is no obstacles.

Possible modes of travel by cognitively impaired when navigating in outdoors are driving and walking. Both of these modes of travel share the same routing preferences that include no unfamiliar areas, no congestions, least number of turns, and no unsafe neighborhoods. The only mode of travel by cognitively impaired when navigating in indoors is walking with such routing preferences as no administrative/maintenance areas and no private spaces.

Possible modes of travel by elderly when navigating in outdoors are driving and walking. One major routing preference by elderly when driving in outdoors is least number of left turns. When walking in outdoors, no steps and no obstacles are the main routing preferences. Walking is the only mode of travel by elderly with avoid steps as the main routing preference. However, as shown in Figure 7.2, the elderly group's navigation requirements and preferences overlap those of the other groups'. For example, walking with walker, which is one possible mode of travel by mobility impaired, could also be required by the elder. Similarly, routing preferences by the elderly group may include those preferred by the other groups.

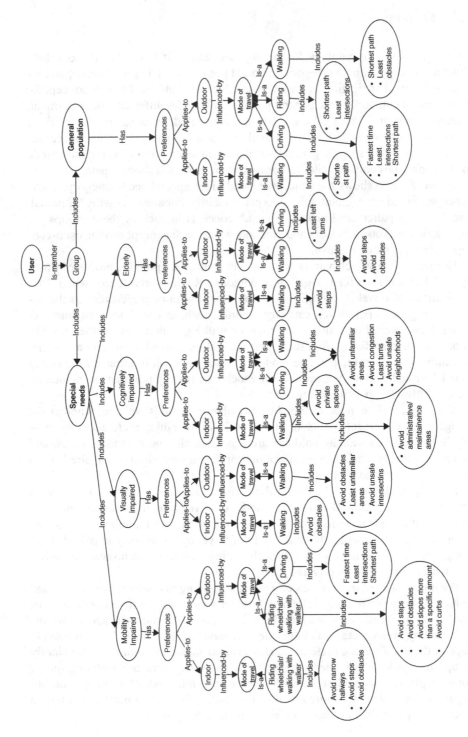

Fig. 7.3 An ontology for anyuser feature of UNAVIT

7.3   Requirements

People with special needs and preferences are faced with unique challenges that impact their mobility. In order to understand such challenges, two types of environmental barriers, permanent and temporary, are defined. Permanent environmental barriers are those barriers that permanently exist in the environment. An example of permanent environmental barriers in outdoors is uneven surface on sidewalks impeding mobility of wheelchair riders. An example of permanent environmental barriers in indoors is narrow hallway segment inhibiting passage by wheelchair riders. Temporary environmental barriers are those barriers that exist only for a duration of time. An example of temporary environmental barriers in outdoors is accumulated snow on a road segment that temporary blocks the road for passage. An example of temporary environmental barriers in indoors is a hallway under construction.

Table 7.1 shows permanent environmental barriers impacting mobility of people both in outdoors and indoors. In outdoors, mobility of pedestrians is horizontal where the environment could be sidewalks, footpaths, and crosswalks. Permanent environmental barriers include width, surface, curb, and slope. Mobility with respect to POIs in outdoors could be horizontal or vertical and the environment could be entrance path or entrance door. Permanent environmental barriers for entrance path include length, automatic/manual door, and slope. Permanent environmental barriers for entrance door include automatic/manual door, width, and depth of opening. In indoors, mobility of pedestrians could be horizontal or vertical. Hallway segments are the environments for horizontal mobility that include such barriers as width, surface, protruding objects, and height. Vertical mobility could be through elevators, which include such barriers as area, buttons, and door, or could be through stairways, which includes such barriers as handrail, width, height of

Table 7.1 Permanent environmental barriers

Navigation	Environment	Mobility	Environment	Barriers
Outdoor	Pedestrian network	Horizontal	Sidewalk Footpath Crosswalk	Width, surface, curb, slope
	POI	Horizontal/ Vertical	Entrance path	Length, slope, automatic/ manual door, slope
			Entrance door	Automatic/manual door, width, depth of opening
Indoor	Pedestrian network	Horizontal	Hallway segments	Width, surface, protruding objects, height
		Vertical	Elevator	Area, buttons, door
			Stairways	Handrail, width, height of steps, depth of steps
	POI	N/A	Restroom	Width, grab bar length, grab bar height, lavatory height, lavatory depth, area
			Drinking fountain	Depth, height, width
			Room	Door, area

Table 7.2 Temporary environmental barriers

Navigation Environment		Mobility	Environment	Barriers
Outdoor	Pedestrian network	Horizontal	Sidewalk Footpath Crosswalk	Natural: snowy/rainy weather, darkness, rush hour Man-made: construction
	POI	Horizontal/ Vertical	Entrance path	Natural: snowy/rainy weather; Man-made: construction
			Entrance door	Under repair
Indoor	Hallway network	Horizontal Vertical	Hallway segments Elevator Stairways	Construction Under repair Under repair
	POI	N/A	Restroom Drinking fountain Room	Under repair Under repair Renovation

steps, and depth of steps. Mobility with respect to POIs in indoors could be affected with such environments as restroom, drinking fountain, and room. Permanent environmental barriers in restrooms include width, grab bar length, grab bar height, lavatory height, lavatory depth, and area, related to drinking fountain include depth, height, and width, and in rooms include door and area.

Table 7.2 shows temporary environmental barriers impacting mobility of people both in outdoors and indoors. In outdoors, mobility of pedestrians is horizontal where the environment could be sidewalks, footpaths, and crosswalks. Temporary environmental barriers could be either natural (e.g., snowy/rainy weather, darkness, and rush hours) or man-made (e.g., construction). Mobility with respect to POIs in outdoors could be horizontal or vertical and the environment could be entrance path or entrance door. Temporary environmental barriers for entrance path could be either natural (e.g., snowy/rainy weather) or man-made (e.g., construction). In indoors, mobility of pedestrians could be horizontal or vertical. Hallway segments are the environments for horizontal mobility that include temporary barriers such as construction. Vertical mobility could be through elevators or stairways which could become barriers temporarily (e.g., during elevator malfunction). Mobility with respect to POIs in indoors could be affected with such environments as restroom, drinking fountain, and room. An example of a temporary environmental barrier in a restroom or related to a drinking fountain is when it is under repair. An example of temporary environmental barriers for a room is when it is under renovation.

Table 7.3 shows the different needs and preferences of users for navigation assistance. In this table, five categories of users are considered: general population, mobility impaired, visually impaired, cognitively impaired, and elderly. To understand navigation assistance for different users, the requirements and preferences of different users for routes in different modes of travel are analyzed.

Table 7.3 Route needs and preferences of different users

Group	Optimal Route Needs			
	Outdoor			Indoor
	Driving	Walking	Riding	
General population	Shortest path; Fastest time; Least intersections; Avoid tolls; Least left turns	Shortest path; Least obstacles	Shortest path; Fastest time; Least intersections; Avoid tolls; Least left turns	Shortest path
Mobility Impaired	Shortest path; Fastest time; Least intersections; Avoid tolls; Least left turns	N/A	Avoid steps; Avoid slopes more than a specific amount; Avoid obstacles; Avoid crowded sidewalks; Avoid narrow sidewalks; Avoid curbs	Avoid steps; Avoid narrow hallways; Avoid obstacles
Visually Impaired	N/A	Avoid obstacles; Least unfamiliar routes; Avoid unsafe intersections	N/A	Avoid obstacles
Cognitively Impaired	Avoid congestion; Least steps/turns; Avoid unsafe neighborhoods; Avoid unfamiliar areas	Most direct; Avoid least unfamiliar roads; Routes passing more landmarks; Avoid congestion; Least steps/turns; Avoid unsafe neighborhoods; Avoid unfamiliar areas; Avoid ill-defined pathways (e.g., grassy areas, walking paths with shared space/ bicycle lanes)	N/A	Avoid private spaces of others; Avoid administrative and maintenance areas; Avoid exiting the premises
Elderly	Least left turns	Most direct; Avoid least unfamiliar roads; Routes with more landmarks; Avoid steps; Avoid slopes more than a specific amount; Avoid obstacles	N/A	Avoid steps

In outdoors, individuals belonging to the general population category may drive cars, walk, or ride bicycles. Routing preferences by individuals in the general population category when driving cars include shortest path, fastest time, least intersections, no tolls, and least left turns; when walking include shortest path and least obstacles; when riding (bicycles) include shortest path, fastest time, least intersections, no tolls, and least left turns. In indoors, walking is the only mode of travel by individuals belonging to the general population category. Routing preference by individuals in the general population category when walking indoors is mainly shortest path.

In outdoors, individuals belonging to the mobility impaired group may drive or ride wheelchairs. Routing needs and preferences by individuals in the mobility impaired group when driving include shortest path, fastest time, least intersections, no tolls, and least left turns; when riding wheelchairs include no steps, no slopes more than a specific amount, no obstacles, no traffic (crowded sidewalks), no narrow sidewalks, and no curbs. In indoors, riding wheelchairs is the only mode of travel by individuals belonging to the mobility impaired group. Routing needs and preferences by individuals belonging to the mobility impaired group include no steps, no narrow hallways, and no obstacles.

In outdoors, walking is the only mode of travel by individuals belonging to the visually impaired group. Routing needs and preferences by individuals belonging to the visually impaired population include no obstacles, least unfamiliar routes, and no unsafe intersections. In indoors, walking is the only model of travel by individuals belonging to the visually impaired group. Routing needs and preferences by individuals belonging to the visually impaired group include no obstacles.

In outdoors, individuals belonging to the cognitively impaired group may drive or walk. Routing needs and preferences by individuals belonging to the cognitively impaired group when driving include no congestions, least steps/turns, no unsafe neighborhoods, and no unfamiliar areas; when walking include most direct, no unfamiliar roads, no congestions, least steps/turns, no unsafe neighborhoods, no unfamiliar areas, no ill-defined pathways (e.g., grassy areas, walking paths with shared space/bicycle lanes) and avoid routes with least recognizable landmarks. In indoors, walking is the only mode of travel by individuals belonging to the cognitively impaired group. Routing needs and preferences of individuals belonging to the cognitively impaired group include no private property and no administrative and maintenance areas.

In outdoors, individuals belonging to the elderly group may drive or walk. Routing needs and preferences by individuals belonging to the elderly group when driving include least left turns; when walking include most direct routes, no unfamiliar roads, routes with more landmarks, no steps, no slopes more than a certain amount, and no obstacles. In indoors, walking is the only mode of travel by individuals belonging to the elderly group. Routing needs and preferences by individuals belonging to the elderly group include no steps.

7.4 Algorithms

Figure 7.4 shows an algorithm that provides navigation assistance to the general population both in outdoors and indoors. In the first step, the algorithm determines whether navigation is in outdoor or indoor. Once it is determined that the navigation is in outdoor, the algorithm determines mode of travel. If mode of travel is driving, then a route using the road network of the navigation environment and by taking user's routing preferences into account is computed. This will be followed by map matching on the road network to continuously track user's location on the computed route. If mode of travel is riding bike, similar to driving as mode of travel, a route using the road network of the navigation environment and by taking user's routing preferences into account is computed. Again, similar to driving as mode of travel, this will be followed by map matching to continuously track user's location on the computed route. If mode of travel is walking, then a route using the sidewalk of the navigation environment and by taking user's routing preferences into account is computed. This will be followed by map matching on the sidewalk to continuously track user's location on the computed route.

Once it is determined that navigation is in indoor and the only mode of travel is walking, the shortest route using the hallway network of the building is computed. This will be followed by map matching on the hallway network to continuously track user's location on the computed route.

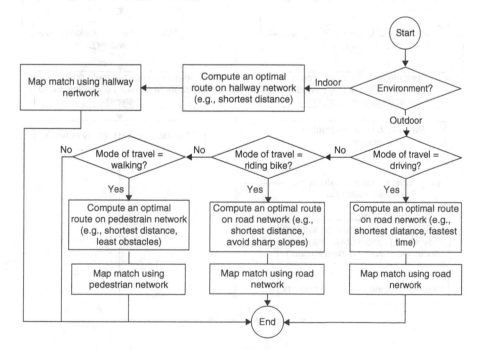

Fig. 7.4 Navigation algorithm for general population

Figure 7.5 shows an algorithm that provides navigation assistance to the mobility impaired both in outdoors and indoors. In the first step, the algorithm determines whether navigation is in outdoor or indoor. Once it is realized that navigation is in outdoor, the algorithm will determine mode of travel. If mode of travel is driving, a route using the road network of the navigation environment and by taking user's routing preferences into account is computed. This will be followed by map match-

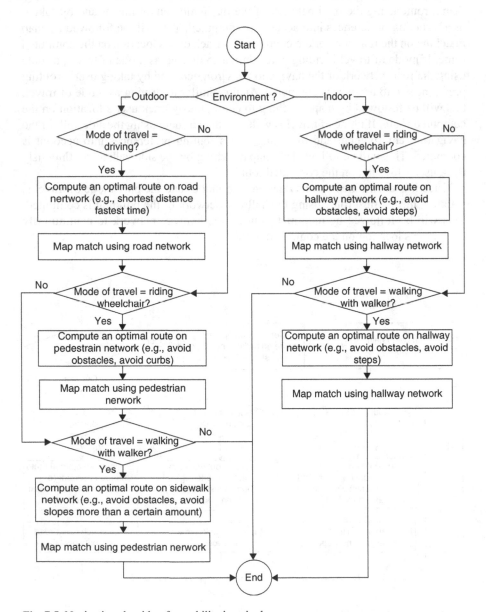

Fig. 7.5 Navigation algorithm for mobility impaired

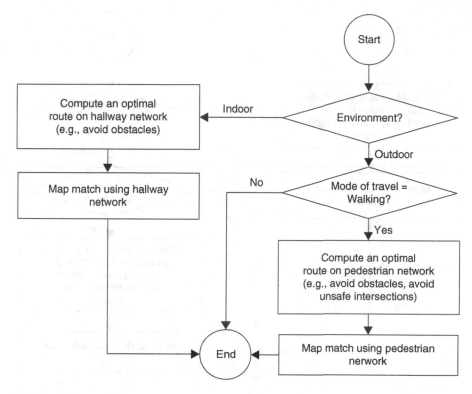

Fig. 7.6 Navigation algorithm for visually impaired

ing on the road network to continuously track user's location on the computed route. If mode of travel is riding wheelchair, a route using the sidewalk network of the navigation environment and by taking user's routing preferences into account is computed. This will be followed by map matching on the sidewalk network to continuously track user's location on the computed route. If mode of travel is walking with walker, similar to riding wheelchair, a route using the sidewalk of the navigation environment and by taking user's routing preferences into account is computed. Again, similar to riding wheelchair, this will be followed by map matching on the sidewalk network to continuously track user's location on the computed route.

Once it is determined that navigation is in indoor, the algorithm will check for mode of travel, which could be either riding wheelchair or walking with walker. In either mode of travel, a route using the hallway network of the building and by taking user's preferences into account is computed. This will be followed by map matching on the hallway network to continuously track user's location on the computed route.

Figure 7.6 shows an algorithm that provides navigation assistance to the visually impaired both in outdoors and indoors. In the first step, the algorithm determines whether navigation is in outdoor or indoor. Once it is determined that navigation is in outdoor, the algorithm proceeds with the only possible mode of travel, which is

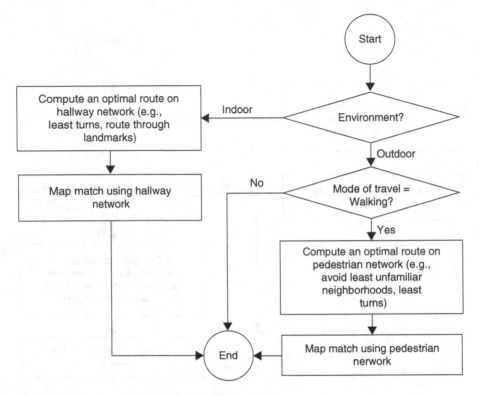

Fig. 7.7 Navigation algorithm for cognitively impaired

walking, and computes a route using the sidewalk of the navigation environment and by taking user's routing preferences into account. This will be followed by map matching on the sidewalk network to continuously track user's location on the computed route.

Once it is determined that navigation is in indoor, with walking as the only mode of travel, a route using the hallway network of the building and by taking user's routing preferences into account is computed. This will be followed by map matching on the hallway network to continuously track user's location on the computed route.

Figure 7.7 shows an algorithm that provides navigation assistance to the cognitively impaired both in outdoors and indoors. In the first step, the algorithm determines whether navigation is in outdoor or indoor. Once it is determined that navigation is in outdoor, the algorithm proceeds with the only possible mode of travel, which is walking, and computes a route using the sidewalk of the navigation environment and by taking user's routing preferences into account. This will be followed by map matching on the sidewalk network to continuously track user's location on the computed route.

Once it is determined that navigation is in indoor, with walking as the only mode of travel, a route using the hallway network of the building and by taking user's rout-

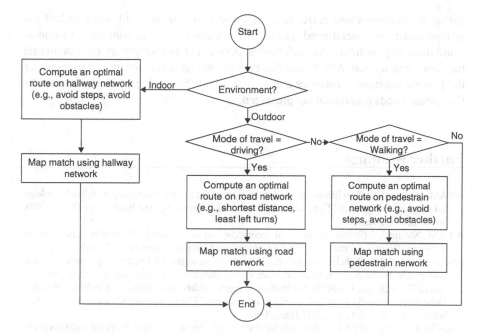

Fig. 7.8 Navigation algorithm for elderly

ing preferences into account is computed. This will be followed by map matching on the hallway network to continuously track user's location on the computed route.

Figure 7.8 shows an algorithm that provides navigation assistance to the elderly both in outdoors and indoors. In the first step, the algorithm determines whether navigation is in outdoor or indoor. Once it is determined that navigation is in outdoor, the algorithm proceeds with the only possible mode of travel, which is walking, and computes a route using the sidewalk of the navigation environment and by taking user's routing preferences into account. This will be followed by map matching on the sidewalk network to continuously track user's location on the computed route.

Once it is determined that navigation is in indoor, with walking as the only mode of travel, a route using the hallway network of the building and by taking user's routing preferences into account is computed. This will be followed by map matching on the hallway network to continuously track user's location on the computed route.

7.5 Summary

In this chapter, the anyuser feature of UNAVIT is discussed. Anyuser in UNAVIT is a reference to personalized navigation assistance. To better understand the anyuser feature of UNAVIT, users are divided into two categories: general population and special needs population. The special needs population is divided into four groups: mobility impaired, visually impaired, cognitively impaired, and elderly. The navigation requirements and preferences of these two categories of users and of the groups under the special needs population are discussed and analyzed. To understand these requirements and preferences, permanent and temporary environmental barriers are described. Algorithms that provide navigation assistance to users under the general population category and to individuals belonging to each group under the special needs population are presented.

Further Readings

Ali AM, Nordin MJ (2009) Indoor navigation to support the blind person using weighted topological map. In: International Conference on Electrical Engineering and Informatics (ICEEI '09), pp 68-72 Malaysia.

Ali AM, Nordin MJ (2009) Vision based reconstruction multi-clouds of scale invariant feature transform features for indoor navigation. Journal of Computer Science 5:948-955.

Al-Rousan M (2005) Webchair: A web-based wireless navigation wheelchair system for people with motor disabilities. International Journal of Computers and Applications 27:274-284.

Amemiya T, Sugiyama H (2009) Navigation in eight cardinal directions with pseudo-attraction force for the visually impaired. In: IEEE International Conference on Systems, Man and Cybernetics (SMC 2009), pp 27-32 Texas, USA.

Andó B, Ascia A (2007) Navigation aids for the visually impaired: From artificial codification to natural sensing. IEEE Instrumentation and Measurement Magazine 10:44-51.

Aono Y, Oichi A, Tadokoro Y (1997) Walking navigation system for the visually impaired using a guide stick. Electrical Engineering in Japan (English translation of Denki Gakkai Ronbunshi) 119:69-79.

Beale L, al e (2006) Mapping for wheelchair users: route navigation in urban spaces. The Cartographic Journal 43:68-81.

Beale L, Field K, Briggs D, Picton P, Matthews H (2006) Mapping for wheelchair users: route navigation in urban spaces. The Cartographic Journal 43:68-681.

Bebek O, Suster MA, Rajgopal S, Fu MJ, Huang X, Çavuolu MC, Young DJ, Mehregany M, Van Den Bogert AJ, Mastrangelo CH (2010) Personal Navigation via High-Resolution Gait-Corrected Inertial Measurement Units. IEEE Transactions on Instrumentation and Measurement.

Bekiaris E, Mousadakou MPA (2007) Elderly and Disabled Travelers Needs in Infomobility Services. In: 4th International Conference on Universal Access in Human-Computer Interaction (UAHCI 2007), pp 853-860 Beijing, China.

Bogdanov I, Tiponut V, Mirsu R (2009) New achievements in assisted movement of visually impaired in outdoor environments. WSEAS Transactions on Circuits and Systems 8:757-768.

Bourbakis N (2008) Sensing surrounding 3-D space for navigation of the blind. IEEE Engineering in Medicine and Biology Magazine 27:49-55.

Calder DJ (2009) Travel Aids For The Blind—The digital ecosystem solution. In: 7th IEEE International Conference on Industrial Informatics (INDIN 2009), pp 149-154 Osaka, Japan.

Chang YJ, Chu YY, Chen CN, Wang TY (2008) Mobile computing for indoor wayfinding based on bluetooth sensors for individuals with cognitive impairments. In: 3rd International Symposium on Wireless Pervasive Computing (ISWPC 2008), pp 623-627 Santorini.

Chang YJ, Wang TY (2010) Indoor wayfinding based on wireless sensor networks for individuals with multiple special needs. Cybernetics and Systems 41:317-333.

Chang YJ, Yan-Ru C, Chang CY, Wang TY (2009) Video prompting and indoor wayfinding based on bluetooth beacons: A case study in supported employment for people with severe mental illness. In: International Conference on Communications and Mobile Computing (CMC '09), vol. 3, pp 137-141.

Chien JC, Lu BY, Lai JS, Luh J, Chong FC, Kuo TS (2010) Electric compass aided global positioning system navigation for powered wheelchairs. Disability and Rehabilitation: Assistive Technology 5:223-229.

Choudhury MH, Barreto A (2003) Design of a multi-sensor sonar system for indoor range measurement as a navigational aid for the blind. Biomedical Sciences Instrumentation 39:30-35.

Cortés U, Urdiales C, Annicchiarico R, Barrué C, Martinez AB, Caltagirone YC (2007) Assistive Wheelchair Navigation: a cognitive view, Advanced Computational Intelligence Paradigms. In: Healthcare, vol. 48 (Yoshida, H. et al., eds), pp 165-187: Springer Berlin / Heidelberg.

Dakopoulos D, Bourbakis NG (2010) Wearable obstacle avoidance electronic travel aids for blind: A survey. IEEE Transactions on Systems, Man and Cybernetics Part C: Applications and Reviews 40:25-35.

Ding D, Parmanto B, Karimi HA, Roongpiboonsopit D, Pramana G, Conahan T, Kasemsuppakorn P (2007) Design Considerations for a Personalized Wheelchair Navigation System. In: the 29th Annual International Conference of the IEEE EMBS Lyon, France.

Dodson AH, Moon GV, Moore T, Jones D (1999) Guiding blind pedestrians with a personal navigation system. Journal of Navigation 52:330-341.

Falco JM, Idiago M, Delgado AR, Marco A, Asensio A, Cirujano D (2010) Indoor Navigation Multi-agent System for the Elderly and People with Disabilities. In: 8th International Conference on Practical Applications of Agents and Multiagent Systems, pp 437-442 Salamanca, Spain.

Fezari M (2007) A navigation system for a wheelchair user based on a multi-modal design. In: Proceedings of the IEEE International Conference on Electronics, Circuits, and Systems, pp 479-482 Marrakech, Morocco.

Fickas S, Sohlberg M, Hung P-F (2008) Route-following assistance for travelers with cognitive impairments: A comparison of four prompt modes. International Journal of Human-Computer Studies 66:876-888.

Gaunet F (2006) Verbal guidance rules for a localized wayfinding aid intended for blind-pedestrians in urban areas. Universal Access in the Information Society 4:338-353.

Golledge RG, Loomis JM, Klatzky RL, Flury A, Xiao Li Y (1991) Designing a personal guidance system to aid navigation without sight: progress on the GIS component. International Journal of Geographical Information Systems 5:373-395.

Grasse R (2010) Assisted Navigation for Persons with Reduced Mobility: Path Recognition Through Particle Filtering (Condensation Algorithm). Journal of intelligent & robotic systems.

Hagethorn FN (2008) Creating design guidelines for a navigational aid for mild demented pedestrians. In: Proceeding of European Conference AmI, pp 276-289 Nuremberg, Germany.

Harrison B, Wu H, Marshall A, Yu W (2004) The ENABLED Indoor/Outdoor Navigation Systems for the Blind and Visually Impaired. In: IEEE ICT Brazil.

Hirahara Y, Sakurai Y, Shiidu Y, Yanashima K, Magatani K (2006) Development of the navigation system for the visually impaired by using white cane. Conference proceedings : Annual International Conference of the IEEE Engineering in Medicine and Biology Society IEEE Engineering in Medicine and Biology Society Conference 1:4893-4896.

Hirtle S, Karimi H, Brusilovsky P Collaborative navigational assistance for specialized populations. In: the proceedings of "You Are Here 2: 2nd Workshop on Spatial Awareness and Geographic Knowledge Acquisition with Small Mobile Devices",Spatial Cognition 2010, Mt. Hood, Oregon, 2010.

Holone H, Misund G (2008) People Helping Computers Helping People: Navigation For People With Mobility Problems By Sharing Accessibility Annotations. In: the International Conference on Computers Helping People with Special Needs (ICCHP'08), vol. 5105/2008, pp 1093-1100: LNCS 5105, Springer, Heidelberg.

Hub A, Diepstraten, J., Ertl, T. (2004) Design and Development of an Indoor Navigation and Object Identification System for the Blind. In: ACM SIGACCESS conference on Computers and accessibility, pp 147-152 Atlanta, GA, USA.

Hub A, Hartter T, Kombrink S, Ertl T (2008) Real and virtual explorations of the environment and interactive tracking of movable objects for the blind on the basis of tactile-acoustical maps and 3D environment models. Disability and rehabilitation Assistive technology 3:57-68.

Hunaiti Z, Garaj V, Balachandran W (2006) A remote vision guidance system for visually impaired pedestrians. Journal of Navigation 59:497-504.

Hunaiti Z, Garaj V, Balachandran W (2009) An assessment of a mobile communication link for a system to navigate visually impaired people. IEEE Transactions on Instrumentation and Measurement 58:3263-3268.

Hunaiti Z, Garaj V, Balachandran W (2009) An assessment of a mobile communication link for a system to navigate visually impaired people. IEEE Transactions on Instrumentation and Measurement 58:3263-3268.

Jacobson RD (1996) Auditory beacons in environment and model: an orientation and mobility development tool for visually impaired people. Swansea Geographer 33:49-66.

Kalia A, Legge G, Roy R, Ogale A (2010) Assessment of Indoor Route-finding Technology for People Who Are Visually Impaired. Journal of Visual Impairment & Blindness, 104:135-147.

Kaluwahandi S, Tadokoro Y (2001) Study of a portable traveling support system using image processing for the visually impaired. Kyokai Joho Imeji Zasshi/Journal of the Institute of Image Information and Television Engineers 55:1499-1505.

Karimanzira D, Otto P, Wernstedt J (2006) Application of machine learning methods to route planning and navigation for disabled people. In: The 25th IASTED International Conference Lanzarote.

Kasemsuppakorn P, Karimi H Data requirements and spatial database for personalized wheelchair navigation. In: 2nd International Convention on Rehabilitation Engineering & Assistive Technology, Bangkok, Thailand.

Kasemsuppakorn P, Karimi HA (2009) Personalised routing for Wheelchair Navigation. Journal of Location Based Services 3:24-54.

Kawamura T, Umezu K, Ohsuga A (2010) Mobile Navigation System for the Elderly—Preliminary Experiment and Evaluation. In: Proceeding of 5th International Conference, UIC 2008, pp 578-590 Oslo, Norway.

Kim L, Park S, Lee S, Ha S (2009) An electronic traveler aid for the blind using multiple range sensors. IEICE Electronics Express 6:794-799.

Klatzky RL, Marston JR, Giudice NA, Golledge RG, Loomis JM (2006) Cognitive load of navigating without vision when guided by virtual sound versus spatial language. Journal of Experimental Psychology: Applied 12:223-232.

Ladetto Q, Merminod B (2002) In step with INS: Navigation for the blind, tracking emergency crews. GPS World 13:30-38.

Lin PC, Chien LW (2010) The effects of gender differences on operational performance and satisfaction with car navigation systems. International Journal of Human Computer Studies.

Liu XF, Ma JG, Cen M (2008) Analysis of Z axis moving state of three-axis photoelectric tracking system in blind region of horizontal photoelectric tracking system. Guangzi Xuebao/Acta Photonica Sinica 37:2067-2071.

Liu Y-C (2003) Effects of using head-up display in automobile context on attention demand and driving performance. Displays 24:157-165.

Loomis JM, Marston JR, Golledge RG, Klatzky RL (2005) Personal guidance system for people with visual impairment: A comparison of spatial displays for route guidance. Journal of Visual Impairment and Blindness 99:219-232.

Mahmud AA, Mubin O, Shahid S (2009) User experience with in-car GPS navigation systems: Comparing the young and elderly drivers.

Marco A, Casas R, Falco J, Gracia H, Artigas JI, Roy A (2008) Location-based services for elderly and disabled people. Computer Communications 31:1055-1066.

Matthews H, al. e (2003) Modelling access with GIS in urban systems (MAGUS): capturing the experience of wheelchair users. Area 35:34-45.

Nordin MJ, Ali AM (2009) Indoor navigation and localization for visually impaired people using weighted topological map. Journal of Computer Science 5:883-889.

Öktem R, Aydin E (2010) An RFID based indoor tracking method for navigating visually impaired people. Turkish Journal of Electrical Engineering and Computer Sciences 18:185-196.

Oomes AH, Bojic M, Bazen G (2009) Supporting Cognitive Collage Creation for Pedestrian Navigation. In: Proceedings of the 8th International Conference on Engineering Psychology and Cognitive Ergonomics: Held as Part of HCI International 2009, pp 111-119 San Diego, CA.

Pereira M, Bruyas MP, Simóes A (2010) Are elderly drivers more at riskwhen interacting with morethan one in-vehicle system simultaneously? Travail Humain 73:53-73.

Petrie H, Johnson V, Strothotte T, Raab A, Fritz S, Michel R (1996) MOBIC : Designing a travel aid for blind and elderly people. Journal of Navigation 49:45-52.

Postolache O (2009) UbiSmartwheel—A ubiquitous system with unobtrusive services embedded on a wheelchair. In: The 3rd International Conference on PErvasive Technologies Related to Assistive Environments (PETRA'09) Corfu, Greece.

Prabu I (2008) GPS and GIS for the blind: A navigation kit for the visually impaired. GIM International 22:40-41.

Punwilai J, Noji T, Kitamura H (2009) The design of a voice navigation system for the blind in negative feelings environment. In: 9th International Symposium on Communications and Information Technology (ISCIT 2009), pp 53-58 Icheon, South Korea.

Ran L, Helal S, Moore S (2004) Drishti: An Integrated Indoor/Outdoor Blind Navigation System and Service. In: The Second Annual Conference on Pervasive Computing and Communications, pp 23-30 Orlando, FL, USA.

Rashid O, Dunbar A, Fisher S, Rutherford J (2010) Users helping users: User generated content to assist wheelchair users in an urban environment. pp 213-219.

Ren M, Karimi H (2009) A chain-code-based map matching algorithm for wheelchair navigation. Transactions in GIS 13 (2):197-214.

Ren M, Karimi H (2009) A hidden Markov model-based map-matching algorithm for wheelchair navigation. The Journal of Navigation 62 (3):383-395.

Riehle TH, Lichter P, Giudice NA (2008) An indoor navigation system to support the visually impaired. Conference proceedings : Annual International Conference of the IEEE Engineering in Medicine and Biology Society IEEE Engineering in Medicine and Biology Society Conference 2008:4435-4438.

Sáenz M (2009) Indoor position and orientation for the blind. In: Proceedings of the 5th International Conference on Universal Access in Human-Computer Interaction, pp 236-245 San Diego, CA, USA.

Sáenza M, Sáncheza J (2010) Indoor orientation and mobility for learners who are blind. Annual Review of CyberTherapy and Telemedicine 8:131-134.

Sainarayanan G, Nagarajan R, Yaacob S (2007) Fuzzy image processing scheme for autonomous navigation of human blind. Applied Soft Computing Journal 7:257-264.

Sánchez J (2009) Mobile audio navigation interfaces for the blind. In: Proceedings of 5th International Conference (UAHCI 2009) Held as Part of HCI International 2009, pp 402-411 San Diego, CA, USA.

Sánchez J, Sáenz M (2010) Metro navigation for the blind. Computers and Education.

Scherlen AC, Dumas JC, Guedj B, Vignot A (2007) "RecognizeCane" : The new concept of a cane which recognizes the most common objects and safety clues. Conference proceedings : Annual International Conference of the IEEE Engineering in Medicine and Biology Society IEEE Engineering in Medicine and Biology Society Conference 2007:6357-6360.

Seto FT (2009) A navigation system for the visually impaired using colored navigation lines and RFID tags. Conference proceedings : Annual International Conference of the IEEE Engineering in Medicine and Biology Society IEEE Engineering in Medicine and Biology Society Conference 2009:831-834.

Sheehan B, Burton E, Mitchell L (2006) Outdoor wayfinding in dementia. Sage Publications 5:271-281.

Shen H, Chan KY, Coughlan J, Brabyn J (2008) A mobile phone system to find crosswalks for visually impaired pedestrians. Technology and Disability 20:217-224.

Shiizu Y, Hirahara Y, Yanashima K, Magatani K (2007) The development of a white cane which navigates the visually impaired. Conference proceedings : Annual International Conference of the IEEE Engineering in Medicine and Biology Society IEEE Engineering in Medicine and Biology Society Conference 2007:5005-5008.

Shoval S, Borenstein J, Koren Y (1998) Auditory guidance with the navbelt-a computerized travel aid for the blind. IEEE Transactions on Systems, Man and Cybernetics Part C: Applications and Reviews 28:459-467.

Shoval S, Borenstein J, Koren Y (1998) The NavBelt—A computerized travel aid for the blind based on mobile robotics technology. IEEE Transactions on Biomedical Engineering 45:1376-1386.

Simpson R, LoPresti E, Hayashi S, Guo S, Ding D, Ammer W, Sharma V, Cooper R (2005) A prototype power assist wheelchair that provides for obstacle detection and avoidance for those with visual impairments. Journal of NeuroEngineering and Rehabilitation 2.

Sobek AD, Miller HJ (2006) U-access: A web-based syastem for routing pedestrians of differing abilities. Journal of Geographical Systems 8:269-287.

Stankiewicz BJ, Cassandra AR, McCabe MR, Weathers W (2007) Development and Evaluation of a Bayesian Low-Vision Navigation Aid. IEEE Transactions on Systems, Man, and Cybernetics—Part A: Systems and Humans 37:970-983.

Swobodzinski M, Raubal M (2009) An indoor routing algorithm for the blind: Development and comparison to a routing algorithm for the sighted. International Journal of Geographical Information Science 23:1315-1343.

Taha T (2010) POMDP-based long-term user intention prediction for wheelchair navigation. In: Proceedings of the 2008 IEEE International Conference on Robotics and Automation, pp 3920-3925 Pasadena, CA, USA.

Takatori N, Nojima K, Matsumoto M, Yanashima K, Magatani K (2006) Development of voice navigation system for the visually impaired by using IC tags. Conference proceedings : Annual International Conference of the IEEE Engineering in Medicine and Biology Society IEEE Engineering in Medicine and Biology Society Conference 1:5181-5184.

Tao Y, Wang T, Wei H, Chen D (2010) A navigation method based on POMDP for smart wheelchair in uncertain environments. High Technology Letters 16:164-170.

Thapar N, Warner G, Drainoni M, Williams R, Ditchfield H (2004) A pilot of study of functional access to public buildings and facilities for persons with impairments. Disability and Rehabilitation 26:280-289.

Urdiales C (2009) Adaptive collaborative assistance for wheelchair driving via CBR learning. In: IEEE International Conference on Rehabilitation Robotics (ICORR 2009), pp 731-736 Kyoto, Japan.

Urdiales C (2009) A metrics review for performance evaluation on assisted wheelchair navigation. In: Lecture notes in computer science, pp 1145-1152.

Vanhooydonck D, Demeester E, Hüntemann A, Philips J, Vanacker G, Van Brussel H, Nuttin M (2010) Adaptable navigational assistance for intelligent wheelchairs by means of an implicit personalized user model. Robotics and Autonomous Systems 58:963-977.

Völkel T, Gerhard G (2007) A New Approach for Pedestrian Navigation for Mobility Impaired Users Based on Multimodal Annotation of Geographical Data. In: Proceeding of 4th International Conference on Universal Access in Human-Computer Interaction (UAHCI 2007) Held as Part of HCI International 2007, pp 575-584 Beijing, China.

Völkel T, Weber G (2008) RouteCheckr: Personalized multicriteria routing for mobility impaired pedestrians. In: The 10th International ACM SIGACCESS Conference on Computers and Accessibility (ASSETS'08), pp 185-192.

Wada C (2009) Basic study on presenting distance information to the blind for navigation. In: 2009 Fourth International Conference on Innovative Computing, Information and Control Kaohsiung, Taiwan

Yang S, Mackworth AK (2007) Route Planning and Scheduling for Wheelchair Users. In: the Festival of International Conferences on Caregiving, Disability, Aging and Technology (FICCDAT) Toronto, Canada.

Zeng Q (2007) Evaluation of the collaborative wheelchair assistant system. In: IEEE 10th International Conference on Rehabilitation Robotics (ICORR'07), pp 601-608 Noordwijk, ëthe Beach of Amsterdamí, Netherlands.

Zeng Q (2008) User evaluation of a collaborative wheelchair system. In: Proceedings of the 30th Annual International Conference of the IEEE Engineering in Medicine and Biology Society (EMBS'08), pp 1956-1960 Vancouver, British Columbia, Canada.

Zhang J, Ong SK, Nee AYC (2009) Design and Development of a Navigation Assistance System for Visually Impaired Individuals. In: Proceedings of the 3rd International Convention on Rehabilitation Engineering & Assistive Technology (ICREAT '09), pp 1-4 Singapore.

Chapter 8
Social Navigation Networks

8.1 Introduction

In previous chapters, navigation technology was the focus. In Chapter 1, evolution
of navigation technology, its current trend, and its future direction were discussed.
Chapters 2 and 3 provided details about navigation technology for outdoor navigation and for indoor navigation, respectively. In Chapter 4, the emerging trend (i.e.,
UNAVIT) was discussed. Chapters 5, 6, and 7 were devoted to the three main features
of UNAVIT (i.e., anywhere navigation, anytime navigation, and anyuser navigation),
respectively. In this chapter, a new and emerging approach for navigation assistance
is discussed.

This new approach is called Social Navigation Networks (SoNavNets) and is
made possible primarily through Web 2.0 technology. Social networking has recently
gained the attention of developers who build social networks utilizing Web 2.0 technology as the underlying infrastructure and has become popular by users who utilize
social networks for personal and professional activities. In this chapter, SoNavNets,
related technologies and systems, and a prototype SoNavNet are discussed.

8.2 Location-Based Social Networks

One principal difference between Web 1.0 and Web 2.0 is that the former is primarily for information retrieving whereas the latter makes user participation in commu-
nities and collaboration with others possible. Today, there are many different social
networks and membership in them is continually growing.

The popularity of online social networks and Location-Based Services (LBSs)
has paved the way for Location-Based Social Networks (LBSNs) where members
of the social network share and exchange LBS information with each other. The
core information shared and exchanged in LBSNs is on or related to location of objects and people. Mobility is another important feature of LBSNs where locations of
objects (e.g., cars) and locations of people at different times are analyzed and taken
into consideration for decision making. Smartphones are increasingly becoming

H. A. Karimi, *Universal Navigation on Smartphones,*
DOI 10.1007/978-1-4419-7741-0_8, © Springer Science+Business Media, LLC 2011

dominant in LBSNs where they can be used for a variety of activities, but mainly for determining on-demand locations of objects and people.

SoNavNets are specialized LBSNs devoted to navigation information sharing and exchange. While LBSNs and some of the current online social networks provide information related to navigation such as POIs, currently, as of this writing, there is no single online social network whose main purpose is navigation information sharing and exchange among its members. Navigation information in SoNavNets, as a medium for navigation information sharing and exchange, mainly include POIs, routes, and directions. SoNavNets' members would have two modes of interaction: request and recommend. Figure 8.1 shows the information content that SoNavNets store and share. POIs are geotagged (i.e., provided with coordinates) locations of interest, either in outdoors or indoors, where details about them are described. An example POI in outdoors is location of a restaurant with a description of types of food, service, prices, etc. An example POI in indoors is location of a restroom in a building that is accessible to wheelchair riders along with information about the accessories it supports. Information on routes/directions could be for outdoors or indoors. In outdoors, routes/directions are geotagged as lines and/or directions (sequence of instructions on road segments of routes) which could be for driving on roads or for walking or riding wheelchairs on sidewalks. In indoors, people walk or ride wheelchairs on hallways. Modes of interaction in SoNavNets could be request or recommend. Members can recommend their experience about POIs (e.g., suggest POIs) and routes/directions (e.g., suggest routes/directions) to other members. Members can request POIs or routes/directions recommended by other members stored in SoNavNets. In the case of POI request, members look for a POI among all POIs recommended by other members that meets their needs. In the case of route/direction request, members look for a route/direction among all routes/directions recommended by other members that meets their needs.

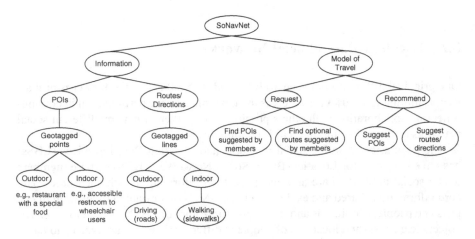

Fig. 8.1 Information content and modes of interaction in SoNavNets

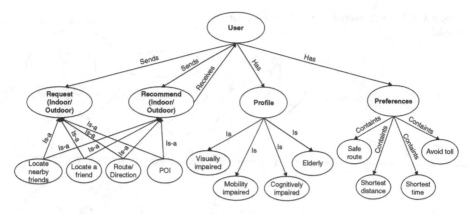

Fig. 8.2 An ontology for SoNavNets

8.3 Ontology

Figure 8.2 shows an ontology for SoNavNets. By joining SoNavNets as a new member, the user provides profile and preferences. In the profile, members can provide their limitations with respect to mobility visual and cognitive impairments. Members can also indicate their preferences for POIs and routes/directions. Similar to navigation technology, routing preferences include shortest distance, fastest time, safest route, or no tolls. The profile and preferences will be used in recommending suitable POIs and routes/directions to the members. Members can recommend POIs, routes, directions, and locate a friend or locate nearby friends. All recommended information can be requested by members, as shown in Figure 8.2.

8.4 Network

Social networks are typically modeled as graphs where nodes are members and links represent interaction among members. Social networks are either general (i.e., their models are not based on a particular pattern) or specialized (i.e., their models are based on specific patterns of information flow and interaction among members). SoNavNets are specialized networks, as shown in Figure 8.3, since they facilitate navigation information sharing and exchange. The overall network is a hyper-network where each node is a sub-network of people who typically share navigation information with each other and the links indicate that members of each group (sub-network) may interact with members of other groups. In other words, the sub-networks, at the nodes of the hyper-network, are strong links and the links among sub-networks are weak links with respect to navigation information sharing and exchange.

Fig. 8.3 SoNavNets overall model

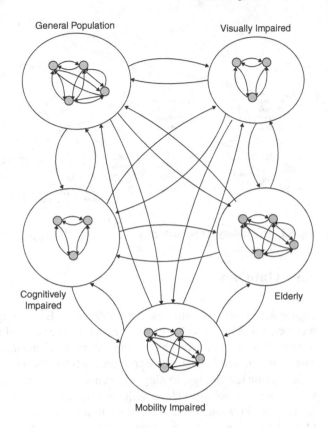

The five groups of users who could share navigation information are general population, mobility impaired, visually impaired, cognitively impaired, and elderly. Each group is expected to have a certain pattern of requesting and recommending navigation information. Whether the patterns among the five groups are similar or not is a topic of further research as SoNavNets are new and require some time before their detailed characteristics can be realized. Furthermore, as shown in Figure 8.3, members of each group can also be members of other groups and can request and recommend navigation information from and to members in other groups, though their interaction may be infrequent, thus weak links.

8.5 Algorithms

Figure 8.4 shows the main functions in SoNavNets which are set profile, messaging, request, and recommend. Each new member will utilize the "set profile" function to include their profile into SoNavNets which includes special needs and preferences for navigation information. Through the "messaging" function, mem-

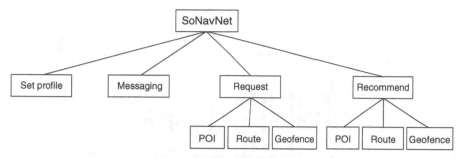

Fig. 8.4 Main functions in SoNavNets

bers of SoNavNets can communicate with all or a selected list of members for the purpose of sharing navigation information. Through the "request" function, members can submit queries about POIs, routes, and directions. Members can use the "recommend" function to suggest new POIs, routes, and directions. Once a request is submitted to SoNavNets, information on member's profile will be used to search through recommended information to find a close match.

Figure 8.5 shows an algorithm for the request function in SoNavNets. Upon submission of a request, the group (e.g., mobility impaired) in which the member would like to search navigation information is determined either through member's profile or member's request. The submitted query will be checked for the type of navigation information required. If the request is a POI, then the preferences of the member, obtained through the profile information, will be matched against the recommended POIs by members of the same group. If a recommended POI is matched with member's preferences, it will be presented to the member, otherwise, other groups in the SoNavNet will be searched for a match. If a match through other groups is found, the matched POI will be presented to the member, otherwise, the member will be notified that no matched POI could be found in the SoNavNet.

If the request is a route, then the preferences of the member, obtained through member's profile or request, will be matched against the recommended routes by members of the same group. If a recommended route is matched with a member's preferences, it will be presented to the member, otherwise, other groups in the SoNavNet will be searched for a match. If a match through other groups is found, the matched route will be presented to the member, otherwise, the member will be notified that no matched route could be found in the SoNavNet.

If the request is a direction, then the preferences of the member, obtained through member's profile or request, will be matched against the recommended directions by members of the same group. If a recommended direction is matched with member's preferences, it will be presented to the member, otherwise, other groups in the SoNavNet will be searched for a match. If a match through other groups is found, the matched direction will be presented to the member, otherwise, the member will be notified that no matched direction could be found in the SoNavNet.

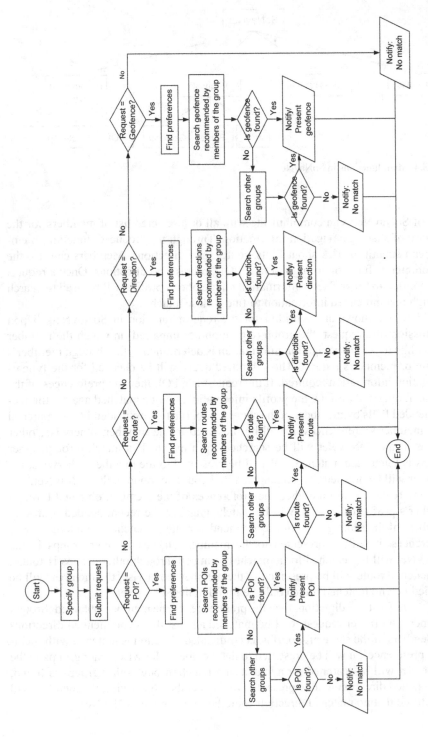

Fig. 8.5 An algorithm for request function in SoNavNets

If the request is a geofence, then the preferences of the member, obtained through member's profile or request, will be matched against the recommended geofences by members of the same group. If a recommended geofence is matched with member's preferences, it will be presented to the member, otherwise, other groups in the SoNavNet will be searched for a match. If a match through other groups is found, the matched geofence will be presented to the member, otherwise, the member will be notified that no matched geofence could be found in the SoNavNet.

Figure 8.6 shows an algorithm for the recommend function in SoNavNets. Upon submission of a recommendation, the member can select a group for which the recommendation is most appropriate. If the recommendation is a POI, the member can geotag the POI, annotate it, and include information about the experience along with a score for the experience. If the recommendation is a route, the member can sketch the route, annotate it, and include information about the experience along with a score for the experience. If the recommendation is a direction, the member can include (e.g., sketch) the direction, annotate it, and include information about the experience along with a score for the experience. If the recommendation is a

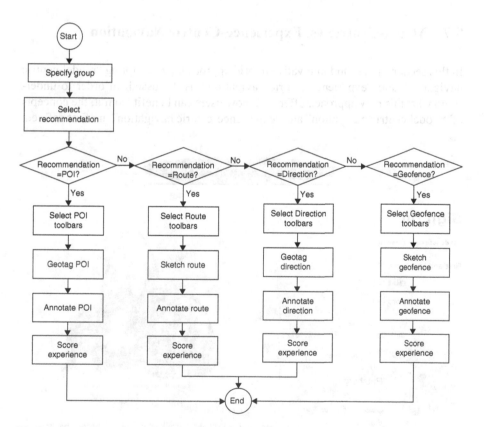

Fig. 8.6 An algorithm for recommend function in SoNavNets

geofence, the member can include sketch the geofence, annotate it, and include information about the experience along with a score for the experience.

8.6 SoNavNet

The researchers in the Geoinformatics Laboratory at the University of Pittsburgh have developed a prototype SoNavNet under the direction of the author of this book. In its current version, the prototype SoNavNet features most of the specifications discussed above. Figures 8.7 through 8.11 illustrate sample screenshots of the prototype SoNavNet. Figure 8.7 shows the main page where new members can sign up for the system. Figure 8.8 shows the page where members can compose new messages to other members. Figure 8.9 shows the page for recommending a new POI (a library in this example). Figure 8.10 shows the page for recommending a route. Figure 8.11 shows the page for recommending a geofence.

8.7 Model-Centric vs. Experience-Centric Navigation

In this section, a new and innovative hybrid approach, i.e., coupling "model-centric navigation" and "experience-centric navigation", is discussed. In order to understand what this new approach offers and how users can benefit from it, the concepts of "model-centric navigation" and "experience-centric navigation" are first defined.

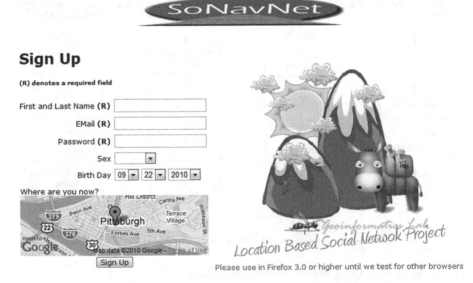

Fig. 8.7 Sign up page in SoNavNet

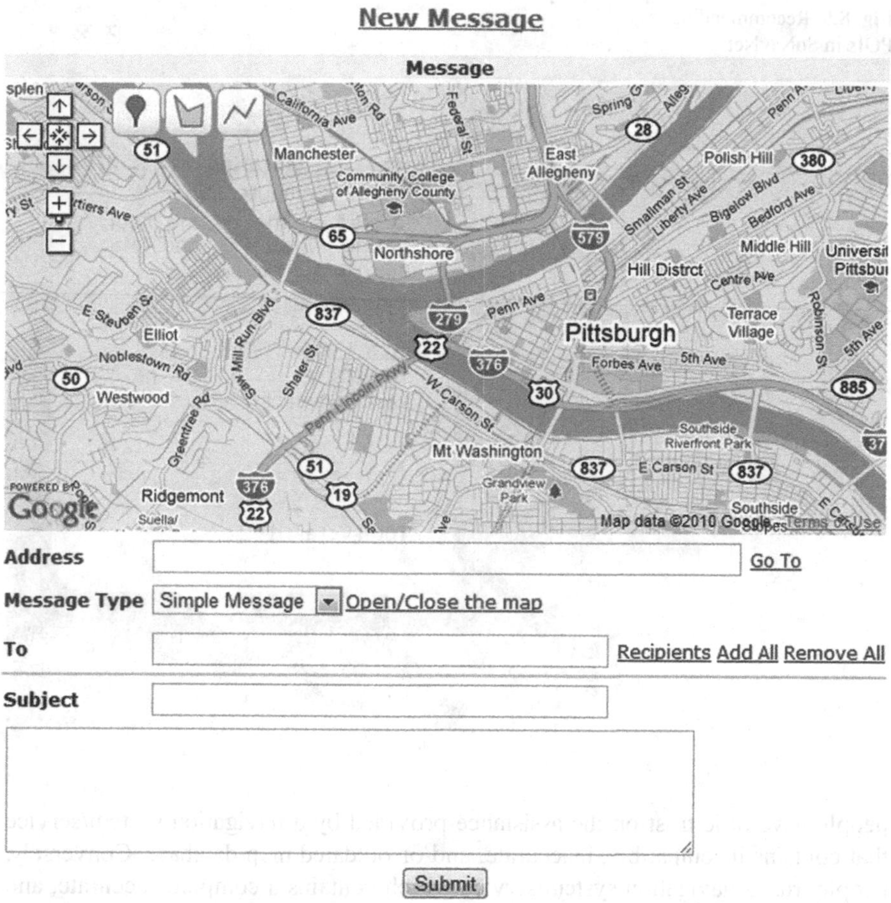

Fig. 8.8 Message page in SoNavNet

Model-centric navigation is a reference to navigation assistance through databases (which are models of the real-world) and computations. Model-centric navigation, which is the foundation of current navigation systems and services, features navigation databases that represent the real-world, in form of road and sidewalk networks in outdoors and of hallway networks in indoors, and navigation computations, such as map matching and routing. Experience-centric navigation is a reference to navigation assistance through sharing experiences of people on POIs, routes, and directions. However, model-centric navigation and experience-centric navigation differ in how they provide navigation assistance. In model-centric navigation, assistance is through computations which include tracking. In experience-centric navigation, assistance is through recommendations and does not include real-time operations.

In model-centric navigation, people tend to rely on navigation assistance that they receive through navigation systems/services if they trust their results. Figure 8.12 shows issues affecting trust on navigation systems/services. As shown in this figure,

Fig. 8.9 Recommending
POIs in SoNavNet

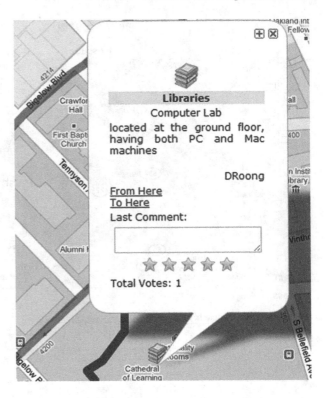

Fig. 8.9 Recommending
POIs in SoNavNet

people have little trust on the assistance provided by a navigation system/service that contains incompatible, inaccurate, and/or outdated map database. Conversely, people trust a navigation system/service which contains a complete, accurate, and up-to-date map database, leading to increase reliance on the results provided by the system/service.

The element of trust plays a similar role in experience-centric navigation. Figure 8.13 shows issues affecting trust on social navigation networks. As shown in this figure, people have little trust on the results provided by a social navigation network that contains inaccurate, non-relevant, and non-personalized recommendations. Conversely, people trust a SoNavNet whose recommendation is accurate, relevant, and personalized, leading to increase reliance on the results provided by the network.

Table 8.1 shows the requirements of model-centric navigation from a data perspective (network) for the purpose of computation. This table summarizes the navigation requirements of people with special needs and preferences. In outdoor navigation systems/services, both points and lines are represented in the vector data model. Geometry of lines composes coordinates of start and end points of each road/sidewalk segment and coordinates of points forming shape of the road/ sidewalk segment. Topology of lines contains connectivity of road/sidewalk segments through intersections. Attributes include common properties, such as name

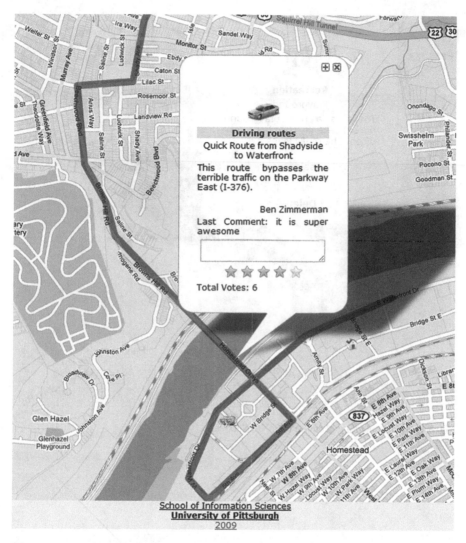

Fig. 8.10 Recommending routes in SoNavNet

and length, static accessibility, such as width, slope, step, and surface, and dynamic accessibility which could be natural, such as surface condition due to snow/rain, or man-made, such as construction and congestion. Geometry of points composes coordinates of intersections and POIs.

In indoor navigation systems/services, both points and lines are represented in the vector data model. Geometry of lines composes coordinates of start and end points of each hallway segment and coordinates of points forming shape of the hallway segment. Topology of lines contains connectivity of hallway segments. Attributes include common properties, such as length, static accessibility, such as width,

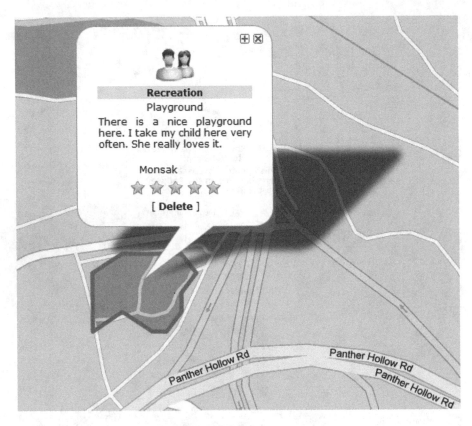

Fig. 8.11 Recommending geofence in SoNavNet

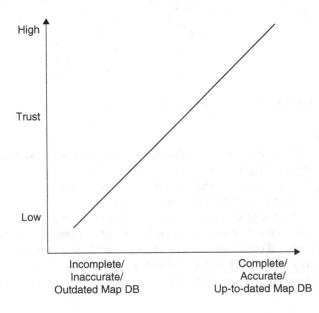

Fig. 8.12 Issues affecting
trust of people on navigation
services

Fig. 8.13 Issues affecting trust of people on social navigation networks

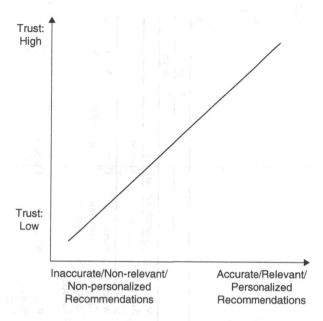

slope, step, and surface, and dynamic accessibility which could be man-made, such as construction. Geometry of points composes coordinates of decision points and POIs.

Table 8.2 shows the requirements of model-centric navigation from a data perspective (point) for the purpose of computation. This table summarizes the navigation requirements of people with special needs. When navigation is in outdoor, point geometry composes coordinates of intersections and POIs. Topology of points contains connectivity of intersections through road/sidewalk segments. Attributes include common propertie, such as name and type (e.g., restaurant, bookstore), static accessibility, such as accessible pathway and accessible entrance, natural dynamic accessibility, such as accessible pathway condition due to snow/rain, and man-made dynamic accessibility, such as construction.

When navigation is in indoor, point geometry composes coordinates of decision points and POIs. Point topology contains connectivity of decision points through hallway segments. Attributes include common properties, such as number and type (e.g., room, restroom, drinking fountain), static accessibility, such as door, area, protruding objects (e.g. fire extinguisher), room, width, grab bar, restroom, width of sink, depth of sink, height of sink, height of drinking fountain, and man-made dynamic accessibility, such as construction.

Table 8.3 shows the requirements of experience-centric navigation from a data perspective. Recommendations for both outdoor and indoor navigation include POI (name, address, accessibility) and route/direction (name, accessibility). Member's profile, which is used for both outdoor and indoor navigation, includes group category (i.e., mobility impaired, visually impaired, cognitively impaired, elderly) and preferences.

Table 8.1 Model-Centric navigation: requirements from a data perspective (network)

Location	Data Model	Lines						Points
		Geometry	Topology	Attribute				Geometry
				Common Properties	Static Accessibility	Dynamic Accessibility		
						Natural	Man-Made	
Outdoor	Vector	Coordinates of start and end points; Coordinates of points forming shape of segment	Connectivity of road/sidewalk segments	Name; Length	Width; Slope; Step; Surface	Surface condition due to snow/rain	Construction; Congestion	2D (X,Y) coordinates
Indoor	Vector (optional)	Coordinates of start and end points; Coordinates of points forming shape of segment	Connectivity of hallway segments	Length	Width; Slope; Step; Surface	N/A	Construction	2D (X,Y) coordinates

Table 8.2 Model-Centric navigation: requirements from a data perspective (POI)

Location	Geometry	Topology	Attribute		Dynamic Accessibility	
			Common Properties	Static Accessibility	Natural	Man-Made
Outdoor	2D (X,Y) coordinates	Connectivity to road/sidewalk network	Name ; Type (e.g., restaurant, bookstore)	Pathway accessibility; Entrance accessibility	Pathway conditiondue to snow/rain	Construction
Indoor	2D (X, Y) coordinates	Connectivity to hallway network	Name Type	Door; Area; Protruding objects (e.g, fire extinguisher)	N/A	Construction
			Restroom	Door; Width; Grab bar; Protruding objects (e.g., fire extinguisher)	N/A	Construction
			Drinking fountain	Width of sink; Depth of sink; Height of sink; Height of fountain	N/A	Construction

Table 8.3 Experience-Centric navigation: requirements from a data perspective

Environment	Non-spatial data	
	Recommendations	Profile
Outdoor	POI:name, address, accessibility; Route/ Direction: name, accessibility	Group (mobility impaired, visually impaired, cognitively impaired, elderly); Preferences
Indoor	POI: name, address, accessibility; Route/Direction:accessibility	

The hybrid approach, which is made possible by coupling model-centic navigation and experience-centric navigation, provides a new means in navigation assistance where people could use either model-centric navigation, experience-centric navigation, or both depending on the level of trust. Figure 8.14 shows that as trust on model-centric navigation dwindles, people tend to resort to experience-centric navigation and vice versa (i.e., as trust on experience-centric navigation dwindles, people tend to resort to model-centric navigation). However, since neither model-centric navigation nor experience-centric navigation is perfect, the hybrid approach is seen most practical.

In the hybrid approach, navigation services and SoNavNets are coupled, as shown in Figure 8.15, where navigation services compute functions such as mapping, positioning, geocoding, and map matching and SoNavNets recommend POIs, routes, directions, geofences by members. This figure also shows that SoNavNets can be used to provide recommendations on POIs, routes, directions to navigation services where they can use the recommendations for tracking. With this, navigation services and SoNavNets augment each other.

In Table 8.4, model-centric navigation and experience-centric navigation for outdoors are compared from a data perspective. In model-centric navigation, real-

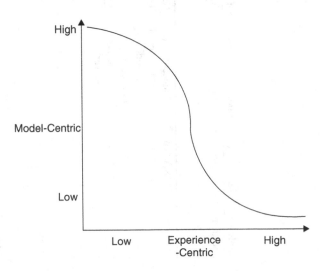

Fig. 8.14 Navigation assistance through the hybrid approach

Fig. 8.15 Navigation services and SoNavNets augmenting each other

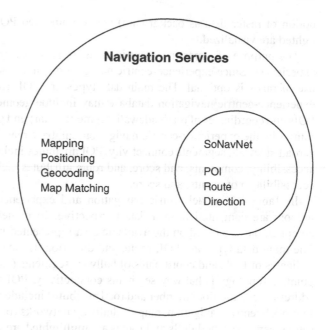

world navigation environments are represented in the vector data model for computations. The main data types are POI, route, and direction. Geometry of data includes coordinates of POIs and coordinates of road/sidewalk segments (start/end points and shape points). Topology is road/sidewalk segments connectivity. POI attributes include name, address, and type, and road attributes include name, length, and speed limit. In model-centric navigation, maps of road/sidewalk networks (with

Table 8.4 Model-centric versus experience-centric navigation for outdoor navigation from a data perspective

Approach		Model-Centric	Experience-Centric
Data Model		Vector; Raster (backdrop)	Optional: vector, raster
Data Type		POI; Route; Direction	POIs; Routes; Direction
Spatial	Attributes	POI:name address, type;	POI: name, address, type, accessibility, comments, score;
		Route: name, length, speed limit	Route: name, type, length, accessibility, comments, score
	Geometry	Coordinates of POIs; Coordinates of road/sidewalk segments	Optional: Coordinates of POIs; Coordinates of road/sidewalk segments
	Topology	Road/Sidewalk segments connectivity	Optional: Road/Sidewalk segments connectivity
Visualization		Map of road/sidewalk networks highlighting POIs and routes	Optional: Map of road/sidewalk networks highlighting POIs and routes
Purpose		Represent navigation environment	Share POI/route/direction experiences

option of raster data as background) where computed POIs and routes are high-lighted are visualized.

The purpose of experience-centric navigation is sharing of POI/route/direction experiences. Since experience-centric navigation is not based on computations, the use of maps is optional. The main data types are POI, route, and direction. The experience-centric navigation database may include geometry, i.e., coordinates of POIs and coordinates of road/sidewalk segments (start/end points and shape points). Similarly, the experience-centric navigation database may include topology which is road/sidewalk segments connectivity. POI attributes include name, address, type, accessibility, comments, and score, and route attributes include name, type, length, accessibility, comments, and score.

In Table 8.5, model-centric navigation and experience-centric navigation for indoors are compared from a data perspective. In model-centric navigation, all computations are based on the available data represented in the vector data model. The main data types are POI, route, and direction. Geometry of data includes co-ordinates of POIs and coordinates of hallway segments (start/end points and shape points). Topology is hallway segments connectivity. POI attributes include name, address, type, and floor number and road attributes include name, width, and length. In model-centric navigation, maps of hallway networks (both vector and raster data sets) where computed POIs and routes are highlighted are visualized.

The purpose of experience-centric navigation is sharing of POIs, routes, and directions. Since experience-centric navigation is not based on computations, the use of maps is optional. The main data types are POI, route, and direction. The experience-centric navigation database may include geometry, i.e., coordinates of POIs and coordinates of hallway segments (start/end points and shape points).

Table 8.5 Model-centric versus experience-centric navigation for indoor navigation from a data perspective

Approach		Model-Centric	Experience-Centric
Data Model		Vector; Raster	Optional: vector, raster
Data Type		POIs; Routes; Direction	POI; Route
Spatial	Attributes	POI: name, type, floor number;	POI: name, address, type, accessibility, comments, score;
		Route: width, length	Route: name, type, width, length, accessibility, comments, score
	Geometry	Coordinates of POIs;	Optional: Coordinates of POIs;
		Coordinates of hallway segments	Coordinates of hallway segments
	Topology	Hallway segments connectivity	Optional: Hallway segments connectivity
Visualization		Map of hallway networks highlighting	Optional: Map of road/sidewalk networks highlighting
		POIs and routes	POIs and routes
Purpose		Represent navigation environment	Share POI/route/direction experiences

Similarly, the experience-centric navigation database may include topology which is hallway segments connectivity. POI attributes include name, address, type, accessibility, comment, and score, and route attributes include name, type, width, length, accessibility, comment, and score.

In Tables 8.6 (a) and 8.6 (b), model-centric navigation and experience-centric navigation are compared from a system perspective. In model-centric navigation, most computations, which are searching or geocoding POIs, finding optimal routes and directions, and tracking, are centralized, either on servers supported by navigation providers or on a mobile device (e.g., smartphone equipped with geo-positioning, wireless communication, and Internet technologies. In model-centric navigation, solutions are 1–1 (i.e., one request results in one solution), computed by the system, whose quality depends on errors in the map database, geo-positioning sensors, geocoding, routing, and direction. In model-centric navigation, POIs, routes, and directions are all computed; POIs are either geocoded or searched in the database. Tracking is accomplished by map matching position data, obtained through

Table 8.6(a) Model-centric versus experience-centric navigation from a system perspective

Approach		Model-Centric	Experience-Centric
Architecture		Centralized	Distributed
Technology		Geopositioning; Wireless communication; Internet	Wireless communication; Internet
Solution		One request, one solution (1-1)	One request, multiple solutions (1-m)
Quality of Solution		Map database, Geopositioning sensors, geocoding, routing, and direction	Members' experiences
Functions	POIs	Geocode; Search DB	Search database
	Routing	Compute	Search database
	Direction	Compute	Search database
	Monitoring	Position data; Map matching	Not possible
	Rerouting	Compute new route/direction	Not possible

Table 8.6(b) Model-centric versus experience-centric navigation from a system perspective

Approach	Model-Centric	Experience-Centric
Routing Criteria	Single criteria	Multiple criteria
Navigation Environment	Map database extent	Geographic areas with recommendations
Mode of Travel	Driving (only in outdoor); Walking; Riding bicycle/wheelchair	Driving (only in outdoor); Walking; Riding bicycle/wheelchair
Platform	Mobile devices (smartphones)	Desktops; Laptops; Smartphones
Purpose	Find POIs; Compute route/direction; Tracking	Share navigation experiences (POIs, routes, directions)

geo-positioning sensors. In case of deviation from a computed route, rerouting occurs, where a new route and direction on it are provided. Typically, single routing criterion is used in model-centric navigation and the navigation environment is only within the extent of the available map database. Navigation assistance with different modes of travel, including driving, walking, and riding bicycles/wheelchairs, can be provided on mobile devices (increasingly on smartphones).

The primary purpose of experience-centric navigation is sharing of navigation experiences (POIs, routes, and directions). In experience-centric navigation, the process is distributed through mobile devices and servers with wireless communication and Internet as the main technologies. Note that geo-positioning sensors are not used in experience-centric navigation as they are in model-centric navigation. In experience-centric navigation, solutions are 1-m (i.e., one request could result in multiple solutions recommended by members), whose quality depends on experiences of members. In experience-centric navigation, POIs, routes, and directions are all searched for recommendations by members in the database and there are no tracking and rerouting. Routing in experience-centric navigation can be based on multiple criteria and the navigation environment is limited to the geographic areas where members' experiences exist. Navigation assistance with different modes of travel, including driving, walking, and riding bicycles/wheelchairs, can be provided on desktops, laptops, and mobile devices (increasingly on smartphones).

In Table 8.7, model-centric navigation and experience-centric navigation are compared from a user perspective. From a user perspective, the purpose of model-centric navigation is navigation for self, whereas the purpose of experience-centric navigation is POI, route, and direction for self and sharing of navigation experiences. In model-centric navigation, users would request tracking, POIs, routes, and directions, whereas in experience-centric navigation, users can request POIs, routes, and directions. While users can recommend POIs, routes, and directions in experience-centric navigation, there is no possibility of recommendation in model-centric navigation. Model-centric navigation can primarily provide the general population with navigation assistance and in special cases it can address the navigation needs of those in such groups as mobility impaired, visually impaired, cognitively impaired, or elderly, whereas experience-centric navigation can assist any individual, whether in the general population or in the special needs groups. One other difference between model-centric navigation and experience-centric navigation from a user perspective is that users in model-centric navigation can be provided with a single routing criterion (e.g., shortest distance, fastest travel time, or least tolls), whereas users in experience-centric navigation can be provided with multiple routing criteria (e.g., fastest travel time and least turns).

Figure 8.16 shows an algorithm for the hybrid approach, coupling navigation services and SoNavNets. As shown in this figure, navigation services obtain position data (through geo-positioning sensors) and match them on road segments or sidewalk segments for navigation in outdoors and on hallways segments for navigation in indoors. Navigation services search POIs and compute routes and directions and geocode by using data in their database. SoNavNets, upon request, find

Table 8.7 Model-centric versus experience-centric navigation from a user perspective

Approach		Model-Centric	Experience-Centric
Request		Tracking POI, route, direction	POI, route, direction
Recommend		Not possible	POI, route, direction
Needs and Preferences	Needs	Primarily general population; one group of special needs (e.g., visually impaired)	General population; special needs groups (mobility impaired, visually impaired, cognitively impaired, elderly)
	Preferences	Single criterion (e.g., shortest distance, fastest travel time, least tolls)	Multiple criteria (e.g., shortest distance and fastest travel time)
Purpose		Navigation for self	POI/routing/direction for self; share navigation experiences

recommended POIs, routes, and directions in their recommendation database. In the hybrid approach, POIs, routes, and directions by both navigation services and SoNavNets are considered and are matched against the needs and preferences of individuals for finding optimal POIs, routes, and directions. Unlike current navigation services that all requests are computed based on the model-centric navigation approach, optimal POIs, routes, and directions in the hybrid approach are decided by considering the results computed through model-centric and the experiences shared through experience-centric navigation. In other words, emerging navigation services will incorporate recommendations by members of the social navigation networks in making navigation decisions, such as optimal routes and directions.

8.8 Summary

The focus of this chapter is online social networking, as a new means for navigation assistance. Social Navigation Networks (SoNavNets) are networks where members can share and exchange navigation information such as POIs, routes, and directions. Characteristics of SoNavNets along with a prototype SoNavNet, developed in the Geoinformatics Laboratory at the University of Pittsburgh under the direction of the author, are discussed. Model-centric navigation, in which real-world navigation environments are represented in form of maps and for which the foundation of current navigation systems and services is provided, and experienced-centric navigation, in which sharing and exchanging navigation information among members are faciliated, are discussed, compared, and contrasted. The hybrid approach (i.e., coupling model-centric navigation and experience-centric navigation) is proposed as a new means for assisting people with their navigation needs and preferences.

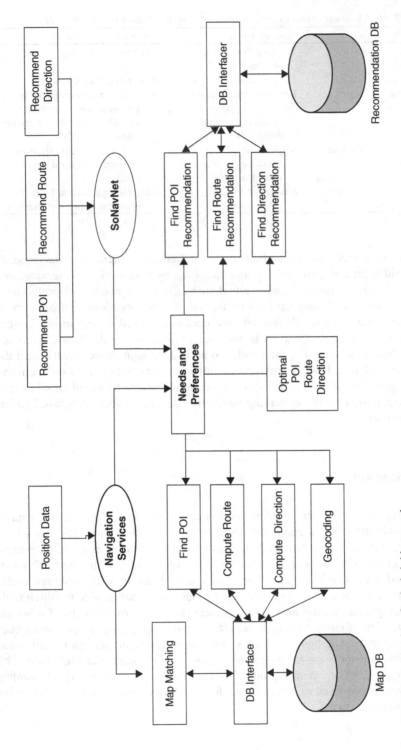

Fig. 8.16 An algorithm for the hybrid approach

Further Readings

Bilton N (2010) Facebook with allow users to share location. http://bits.blogs.nytimes.com/2010/03/09/facebook-will-allow-users-to-share-location/.

Boyd D, Ellison N (2008) Social network sites: Definition, history, and scholarship. Journal of Computer Mediated Communication 13 (1):210-230.

Fusco S, Michael K, Michael M Using a social informatics framework to study the effects of location-based social networking on relationships between people: A review of literature. In: IEEE International Symposium on Technology and Society (ISTAS), Wollongong, NSW, 20 March 2010. IEEE, pp 157-171.

Ghafourian M, Karimi H A, van Roosmalen L Universal navigation through social networking. In: The 13th International Conference on Human-Computer Interaction (HCI 2009), San Diego, California, USA, 19-24 July 2009. Springer, pp 13-22

Holden W (2009) Advertising to fuel mobile social networking growth as UGC revenues reach $7.3bn by 2013. http://juniperresearch.com/shop/viewpressrelease.php?pr=108.

Humphreys L (2008) Mobile social networks and social practice: A case study of Dodgeball. Journal of Computer Mediated Communication 13 (1):341-360.

Karimi H A, Ghafourian M (2009) Universal navigation. GIM International 23:17-19.

Karimi H A, Nwan D, Zimmerman B (2009) Navigation assistance through "Models" or "Experiences"? GIM International December.

Karimi H A, Zimmerman B, Ozcelik A, Roongpiboonsopit D SoNavNet: a framework for social navigation networks. In: the International Workshop on Location Based Social Networks (LBSN'09), Seattle, WA, 3 November 2009. ACM, pp 81-87.

Krishnamurthy B, Wills C Privacy leakage in mobile online social networks. In: Workshop on Online Social Networks, Boston, MA, June 2010. USENIX Association, p 4.

Nakhimovsky Y, Miller A, Dimopoulos T, Siliski M Behind the scenes of google maps navigation: enabling actionable user feedback at scale. In: Proceedings of the 28th of the international conference extended abstracts on Human factors in computing systems, Atlanta, GA, USA, 10-15 April 2010. ACM, pp 3763-3768.

Patel N (2010) Google Buzz takes Location Services to the next level. http://www.engadget.com/2010/02/09/google-buzz-takes-mobile-location-services-to-the-next-level/.

Tsai J, Kelley P, Cranor L, Sadeh N (2010) Location-sharing technologies: Privacy risks and controls. A Journal of Law and Policy for the Information Society 6:119-317.

Twitter (2009) About the Tweet with your location feature. http://twitter.zendesk.com/entries/78525.